Saving the World

Pat McCaw

Illustrations by Caden, Carson, and Wayne Chittick

Enjoy the Adventure!

Pat McCaw

Dedicated to Lily, Sydney, and Ryan - my
support team

Kimmie - who convinced me to write a
book

My parents - who have always believed
in me

Chapter 1

Eleven-year-old Rosie played in the treetops, knew every tree by its leaf, and protected the environment by recycling plastic, planting trees, and composting her brown bananas. Little did Rosie know, a hidden world had noticed her passion and they needed her help.

Rosie rarely spent time inside the house as she explored her backyard. When her parents, Steve and Muriel, decided to move to the countryside months ago, Rosie jumped for joy and planned adventures. Rosie and her dog, Nugget, ran for hours, climbed trees, planted butterfly bushes, or simply sprawled out in the grass watching birds fly overhead. The only bad thing about moving to the country was that her best friend, Lucy, still lived in town. Luckily, Lucy rode the bus home with Rosie after school, and they had a fun afternoon planned.

"I can't believe you like living in the middle of nowhere," Lucy said. "You can't get a Slushee at Rapid Ray's and you have to ride that disgusting, stinky bus." Lucy fluffed her long blonde hair as she took a selfie with her cell phone.

Rosie laughed and said, "What's not to like?" She opened her arms to the sky. "Fresh air and singing birds."

Lucy looked at her manicured pink fingernails and gently kicked away Nugget sniffing at her toes. "Fresh air? I smell pigs." Lucy turned up her nose.

"The neighbors have a farm." Rosie put her hands on her hips. She put up with Lucy's complaints and diva attitude because they had been friends since Unpack your Backpack night in Kindergarten. Lucy had shared her glittered Barbie pencils when Rosie had only plain yellow pencils. Rosie continued, "And I like the bus! I listen to my IPod and watch the rows of corn roll by."

Lucy rolled her eyes. Rosie tugged on Lucy's arm. "I planned something fun today that will convince you that living in the boonies is fun."

Lucy raised her eyebrows and Rosie detected a tiny, itsy-bitsy hint of curiosity. "Fun?"

"A treasure hunt."

Maddy, Rosie's younger sister, barreled onto the porch. She had a yellow bow clipped onto her head full of brown curls, and she wore a dress covered with Minions. "I want to play, I want to play!!! Please, Rosie." Maddy danced back and forth with bare feet on the dirty porch. Maddy hated shoes.

Rosie clenched her teeth. "Maddy, this game is for big kids."

"I just turned six, and Daddy says I'm big."

A BIG pain in the behind, Rosie thought. She started to argue with Maddy when her mom came onto the porch.

"Hi, Lucy. How's your dad doing?" Rosie's mom gave Lucy a sad, empathetic look.

Lucy's mom died one year ago from breast cancer, and her dad still lived in a fog. Recently, his new girlfriend, Judy, sat at Lucy's kitchen table nineteen hours out of the day smoking cigarettes and talking on the phone. Lucy taught herself to cook out of necessity. Judy refused to make anything that didn't come in a box, and Lucy could only tolerate Hamburger Helper so many times before she wanted to crush the adorable glove-with-a-face mascot on the box.

"Hi, Mrs. Montgomery. Dad's okay, I guess." Lucy dropped her head and picked at her perfect pink nails. "I don't see him much. He works late most nights."

"Well, you're welcome to hang out here anytime. I'm glad you three are playing nicely together." Rosie's mom shot her a look and Rosie groaned inside. Maddy was Rosie's shadow forced to follow her everywhere.

Maddy smiled wide and said, "We're going on a treasure hunt, Mom."

Rosie dropped her shoulders. No use fighting it. "Can we borrow some candy for our treasure, Mom?"

The candy bucket was a sacred item in the pantry located high and out of reach. Maddy had an addiction to Dum-Dum suckers and had secretly learned to scale six shelves with her bare feet to sneak the candy when her mom wasn't around. Rosie kept Maddy's secret, but held on to it for later blackmail.

"I suppose a little candy is okay. I'll go get it. Can you believe tomorrow is your last day of school?"

"I can't wait for summer!" Lucy perked up. "One day of torture left and then swimming pools and suntans."

Rosie loved school. "I'm going to miss science class with Mr. Barclay. He does such awesome experiments, but I can't wait to explore the forest trails and see what's out there. Maybe there's a wolf or a badger out back!"

Her mom looked down her nose and pointed her finger. "The forest is off limits. Your dad and I need to check it out before we release you monkeys to the treetops." Her mom stepped back inside to grab the candy treasure after telling Maddy to get her shoes on – again.

Rosie kicked a loose board on the porch. "I want to go into the forest so bad. I don't know what mom and dad are so afraid of?"

Lucy laughed and said, "Spiders, coyotes, rabid bats, goblins."

They laughed.

Rosie planned an elaborate treasure hunt and placed a candy prize at select hiding spots around the yard. The final jackpot would be a bag full of Dum-Dums placed at the trail entrance to the forest. She explained the rules to Lucy and Maddy and mapped it out in her journal. Rosie's journal was a worn, small book kept in her pocket that she sketched or wrote about secrets, interesting facts, future ideas, or anything that popped into her mind. It traveled with her everywhere.

Maddy whined and made Rosie repeat the rules at least seventy-five times before she finally understood. They changed into her dad's camouflage vests and then painted their cheeks with green and brown stripes. Maddy had already kicked off her flip-flops and gripped the grass with her toes. Rosie pulled her long, brown hair into a ponytail and up into a baseball cap and then grabbed her binoculars. Lucy looked hilarious wearing a mini skirt with a camouflage vest and painted face, but she waltzed like a beauty queen down the runway.

"We need to scan the area for our enemies." Getting into character, Rosie climbed the massive oak tree in the center of their backyard. With three quick leaps and then crawling from branch to branch, Rosie perched high in the treetop with her binoculars. Maddy shimmied halfway up behind Rosie and sat on a lower limb.

A tree swing hung from the oak tree, and Lucy plopped down in it as she looked to Rosie up above. "I'll wait down here." Lucy took another selfie and then paused as she looked at her image with the forest behind her. "Doesn't the forest freak you out? How do you know what's out there? There could be a serial killer camped out under a tree."

Rosie had never laughed so hard from the treetop. The forest never scared her and the thought of something bad inside the forest had never crossed her mind. "Very funny, Lucy." She glanced at the map in her journal and knew where the candy was hidden, but wanted to pretend it was a mystery. Their backyard consisted of a thick grass

Treasure Map

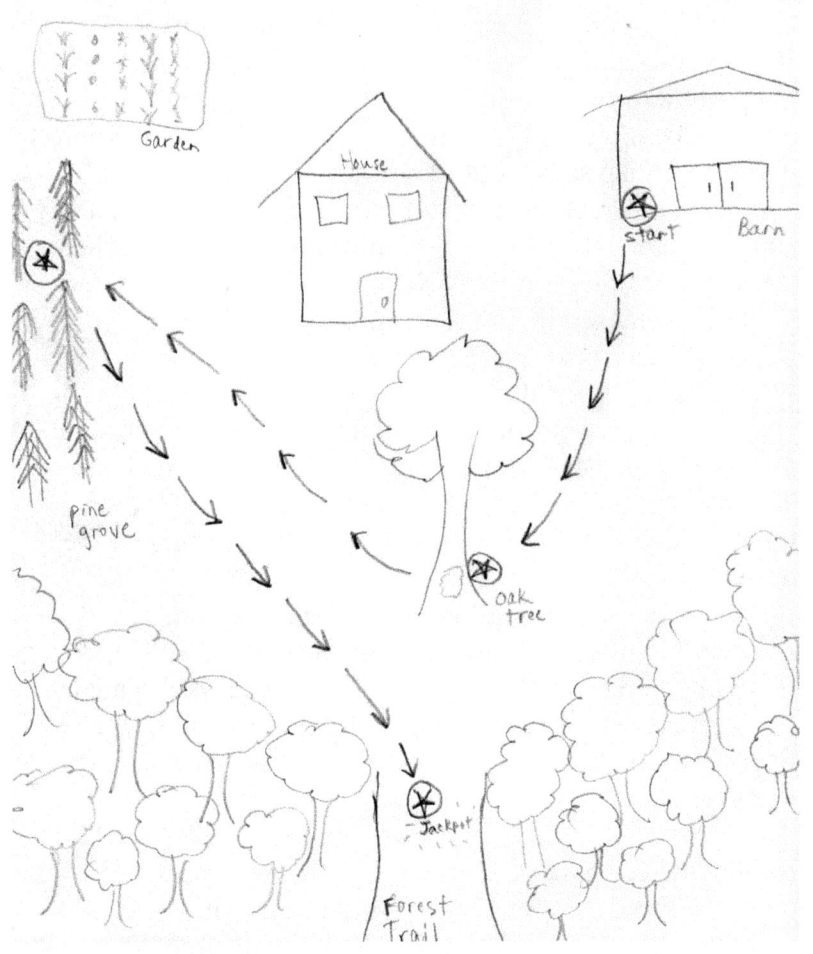

clearing surrounded by dense forest forming the back wall. One side of the yard was lined with a row of pine trees in a grove and her recently planted garden, while on the other side of the yard sat a small barn. Rosie craned her neck to see into the forest, because she had heard that a creek lay hidden in the trees. She dreamed of fishing and wading in its waters, but she couldn't see a thing among the trees. Her stomach flipped with all the possibilities.

"The coast is clear. The treasure hunt is ready." Rosie yelled to Lucy and Maddy while Nugget barked at the base of the tree.

Standing below the tree, they reviewed the treasure map. "The map says we start at the barn."

They crept toward the barn and looked over their shoulder pretending there were thieves trying to steal their treasure. Ducking behind bushes, and sneaking across the yard, they reached their target. After gently opening the barn door, darkness and dust greeted them.

"I don't like the dark, Rosie."

Rosie rarely got scared of anything and was excited to crawl around in darkness, but she flipped on the light with a sigh. "Is that better, Maddy?" Sitting in front of them, was their fishing boat with something sitting on the driver's seat.

"I see something!" Maddy charged ahead.

"What is it?" Lucy said. She inched closer while tiptoeing on the dirt floor to protect her bedazzled sandals.

"Licorice!" Maddy dove into the stash shoved a licorice whip into her mouth. She hummed while she chewed.

After sharing the licorice, the treasure map lead back to the oak tree. They sneaked across the yard and ducked behind shrubs to hide from snipers. They spotted their next treasure tucked inside a knot of the tree, and surprisingly, Lucy ran the fastest to grab it.

5

"Smarties! I love Smarties." Lucy opened a package and popped a pastel orange candy into her mouth.

"These can be our power pellets. They'll give us energy to find the final treasure." Rosie opened a package and ate two candies. She felt powerful immediately.

Next, the map directed them to the pine grove. As they approached the cluster of shaggy trees, Rosie stopped.

"Did you hear that?" Rosie turned toward the forest trail. She swore she had heard rustling in the underbrush. She scanned the trees for a squirrel or raccoon that might be running around.

Maddy and Lucy stared at the trees. Lucy's eyes grew wide. "What did you hear?"

Rosie stood silently and listened. Silence. "Nothing, I guess. Let's keep moving."

Lucy stared at the forest for another minute before she continued to the pine grove. They decided to crawl on their hands and knees, and Maddy enjoyed scooting on the pine needles and getting her dress and knees dirty. Lucy refused to crawl, so she duck-walked to keep out of the filth.

Nugget suddenly barked uncontrollably and ran ahead into the pine trees. Everyone jumped and an eerie silence followed as Nugget suddenly stopped barking, and was nowhere to be seen.

"Rosie, where's Nugget?" Maddy sat up on her heels and tugged on her bottom lip, her telltale sign of fear.

"Nugget!" Rosie called.

Silence.

Lucy yelled louder, "Nugget, come back here right now!"

A rabbit bolted from underneath a pine tree and zipped across their path. Lucy grabbed her chest in fright and Maddy screamed. Nugget barked and chased the rabbit into the yard.

6

Rosie whooped with joy. "That was awesome! Nugget caught our first spy."

Maddy began to cry. "I don't want to play anymore. This is too scary."

Rosie threw her hands in the air. "Maddy, you ruin everything. You can't have any gumballs or Dum-Dums if you quit."

Maddy's lower lip quivered. "I want a Dum-Dum."

"Then keep playing." Rosie had a hard time hiding her frustration. Maddy insisted on joining her but always quit or ran to their mom if she got scared.

They all popped power pellets – or Smarties – into their mouths and continued forward. A colorful box decorated with sequins and gemstones rested against a tree stump. Maddy squealed, "Treasure!"

Gumballs filled the box that Rosie had made in Girls Scouts. They packed their cheeks like chipmunks at an acorn festival.

Rosie whispered, "Now's the most dangerous part of our journey. The final treasure chest is located at the forest entrance. There could be flying monkeys or five-legged bears so we must be prepared." She looked through her binoculars for dramatic effect.

Maddy went pale.

Rosie continued, "Let's throw a dog treat for Nugget next to the treasure chest so that she can chase off the enemy. Maddy, run inside and grab a dog bone."

"Why do I have to?" Maddy stomped her foot.

"Because you're the youngest." Rosie smirked at Lucy.

Maddy's face turned red and she stormed into the house. "I quit!"

"That's a surprise," Rosie yelled to Maddy as she stormed away. Rosie tapped her foot in frustration while Lucy watched the sisterly battle in silence.

Rosie rolled her eyes. "Dang it. I'll grab Nugget's treats and be right back." Rosie sprinted into the house leaving Lucy alone in the yard.

As Rosie rooted in the pantry for a dog treat, her mom cornered her and blocked any escape. "Rosalina Montgomery, why is your sister scared to death? She's babbling about moneys and bears eating up Nugget in the forest."

"It's just a game, Mom. Maddy spoils everything."

"Rosalina, I don't appreciate your tone. Maddy is five years younger than you and I expect you to be nice. Finish your game, because it's almost supper time and I need to run Lucy home."

Rosie shuffled out the back door with and handful of dog treats and a pile of guilt. Lucy stood frozen with both hands over her mouth and wide frightened eyes.

"Lucy, what's wrong?"

Lucy stared at the forest and pointed with a trembling finger.

Rosie shook Lucy's shoulders to make her snap out of it. "Lucy, what happened?"

Lucy mumbled as she focused on the trees. "I heard something."

"What did you hear?" Rosie looked toward the dark forest.

Lucy moved closer to Rosie. "I – I - I don't know. It sounded high-pitched and kind of like a voice."

Rosie had never seen Lucy so scared. Even though she was a diva princess, she could charge into a room without fear. "I'm sorry that I left you outside alone. You're not used to country life and it was probably the cicadas singing." Rosie loved the forest sounds and left her window open every night to listen to the trees and chirping insects.

Lucy snapped. "It wasn't a stupid bug, Rosie. It was not . . . normal."

Rosie's stomach flipped. "Okay."

Lucy sat on the grass twirling her long blonde hair. "Maybe I'm losing my mind."

"How about we finish the game and then go hang out in my room. Mom said she has to run you home pretty soon anyway." Rosie grabbed Nugget's dog bone. "Let's get our Dum-Dums as our reward for being forced to play with Maddy today."

Rosie launched the dog treat through the air and it landed by the forest trail a mere two feet from the Dum-Dum treasure. Nugget sprinted toward the trees. When she approached her bone, she froze. Her ears shot up and she barked ferociously while baring her teeth at the shadows. Nugget took off down the trail.

"Nugget, NO!" Rosie sprinted to the forest, caught up with Nugget, and grabbed her collar. Nugget's body was tense and the hair stood up on her back as she stared into the trees.

That's when Rosie heard it.

A high-pitched noise with a soft rhythm echoed through the trees. Definitely not a cicada! Rosie's heart raced as she tugged on Nugget's collar, grabbed the bag of treasure, and ran toward the house. Tears of fear welled in her eyes as she reached Lucy.

"You heard it, didn't you?" Lucy said with eyes wide. "I knew I wasn't crazy."

"Come on." Rosie tugged Lucy's hand and pulled her into the house and up to her bedroom. Rosie looked over her shoulder and the forest stood dark and mysterious.

When safely on her bed surrounded by the peaceful images of the ocean mural covering her wall, Rosie turned to Lucy. "What the heck was that?"

9

"I have no idea, but I'm never coming to your house again." Lucy grabbed the bag of treasure. "The trees are haunted."

Rosie chuckled. "I doubt that, but I want to know what's out there. Do you think I can convince Mom and Dad to let us hike?"

"Are you crazy! You want to go closer to the creepiness?" Lucy dug her hand inside the treasure bag. "You're that person in scary movies that investigates the sound in the basement. They always get eaten."

Rosie moved to her window and stared out into the dark trees. Her mind raced with a mix of slight fear and anticipation of a secret in her forest. "I need to know what's out there." She replayed the soft rhythms in her head.

"Uh . . . Rosie." Lucy's voice changed and she tossed the bag of treasure onto the desk.

Rosie grabbed the bag and peered inside.

It was empty.

Chapter 2

At the supper table, Rosie picked at her green beans as her mind whirled with the day's adventure. Noises in the forest and stolen candy treasure.

"What's up, Rosey Posey?" Her dad called her Rosie Posey, and insisted on calling Maddy "Fart Blossom." Rosie couldn't break him of this habit.

Rosie knew she couldn't tell her dad about the forest or she would never be able to hike. "Nothing. I'm not very hungry."

Maddy shoveled in her meatloaf as ketchup dripped down her chin.

"Fart Blossom is sure hungry!" Her dad laughed while Rosie cringed.

"I'm going to go study in my room." Rosie pushed away from the table.

"Tomorrow's the last day of school, do you actually have homework?" Her dad questioned and then smiled. "As my future environmentalist, you are probably working on extra credit projects on water purification or solar energy."

Rosie swelled inside. Her parent's understood her passion and never judged her desire to save the world from the people that lived in it. They let her attend Green Camp every summer and their garage had four recycling bins filled to the top. She walked toward the stairs. "Very funny, Dad. I'm just tired." She paused and turned back to her parents. "Do you think we can go hiking in the forest soon?"

Her parents shot glances to one another. Her dad replied, "Soon."

In her room, Rosie opened her window wide and let the night air drift inside. It was getting dark and the forest was black and quiet. She strained her eyes to see – anything.

She hollered out her window, "Anybody out there?"

The crickets chirped and she heard the distant hoot of an owl.

Rosie yelled again, "Who are you?"

A distant sound echoed and then grew from the depths of the forest. A beautiful tune flowed from the trees and into Rosie's window:

> *Fear not young lass, we are your friends*
> *We seek an ally to help defend.*
> *Our world's in danger from an evil foe*
> *Please help us soon, or we're forced to go.*

BANG! BANG! Someone pounded on Rosie's door, and she almost wet her pants.

Her dad peeked his head into her room. "Rosie, are you yelling out the window?"

Rosie turned up the volume on her IPod. "It's just my music, Dad."

"I thought you were studying?" He said with a smile and then got serious. "Maybe we can try out the hiking trails this weekend as a family. That way your mom and I can see what's out there before we feed you to the wolves."

Rosie leaped for joy. "Really! Thanks, Dad."

When the door was closed, Rosie ran back to the window with her heart racing. What did she hear? She strained her ears, but silence filled the forest. The song never returned, so she grabbed her journal and copied down every word she remembered.

> *Fear not young lass, we are your friends*
> *We seek an ally to help defend.*
> *Our world's in danger from an evil foe*
> *Please help us soon, or we're forced to go.*

Rosie heard Maddy singing in her room. She wandered down the hallway and saw Maddy surrounded by her dolls as she changed their diapers and brushed their hair. "What are you doing, Maddy?"

"Getting Jilly and Tinkerbell ready for bed."

Maddy's babies filled her bedroom and Rosie sensed many unblinking eyes staring at her. Dolls sat naked with bows in their hair and Maddy hummed a lullaby while carefully putting pajamas on Baby Jilly.

Did Maddy hear the sounds from the forest? If she asked Maddy and scared her right before bedtime, then her mom would never let Rosie hike for fun on the weekend. Maddy's window was open and the insects chirped outside. Rosie tested her. "Can you hear the crickets chirping tonight?"

Maddy put Baby Tinkerbell's hair in pigtails. "Yep, they're loud."

"Are you excited about the last day of first grade?" Rosie wandered toward Maddy's window and looked outside.

"Yep." Maddy now concentrated on putting pajamas on Baby Tinkerbell.

Rosie gave up and headed toward the door, but then Maddy spoke.

She said, "I sure liked that pretty song."

Rosie froze and turned to Maddy. "What song?"

"The song from outside. I looked out my window and saw something by the oak tree, so I waved." Maddy concentrated on Tinkerbell's buttons and continued. "It went away when you talked to Daddy."

Maddy's tight brown curls flew in all directions as she sat cross-legged in her footed Minion sleeper. She looked so little. Rosie walked back to Maddy's window and her brain exploded with the possibilities. "Weren't you

13

scared?" Maddy couldn't play Rosie's pretend game, but she wasn't fazed by strange sounds from the darkness.

"Nope. They seemed nice."

Rosie's mouth dropped to the floor and she was speechless. What did Maddy mean? Rosie shuffled out of Maddy's room with her mind spinning and crawled into bed. If Maddy saw something in the yard, should she tell their mom and dad? What if it was something dangerous?

Her gut told her that she was safe – and her gut was never wrong.

After her mom kissed her goodnight, Rosie stared at her ceiling imagining all the possibilities.

The last day of school brought many emotions. Something fascinating had happened in the forest, but she hated that it was her last day of school. Mr. Barclay promised a mind-blowing experiment in science class to end the year, so Rosie wore her favorite Einstein T-shirt for the occasion.

Lucy ran up to Rosie in the hallway. "I couldn't sleep last night, and it's all your fault. Your silly treasure hunt."

Rosie laughed and wondered what Lucy would have done if she had heard the forest song last night. "It's a mystery, but I have more to tell you." Rosie looked around and then whispered to Lucy. She showed Lucy her journal with the song from the forest and then told her about Maddy seeing something by the oak tree. Rosie bobbed on her toes as she spoke.

"You're not normal, Rosie. You probably have a murderer in your backyard and you're excited."

Rosie ignored Lucy and continued, "And Dad said we could hike this weekend! You have to come with me."

Despite her lipstick and makeup, Lucy lost all color in her cheeks

They walked into English class and their assignment was to write a short essay on their summer plans. Rosie babbled about cleaning roadways, volunteering at the animal shelter, and hiking in the forest. As they handed in their essays, Lucy flashed her own essay title to Rosie: *A Trip to Disney World.*

Rosie stared in shock. "Are you really going?"

"Of course not. Dad only works or sleeps, while Judy fills me with second-hand smoke." Lucy's eyes softened and she stared at her paper. "But the teacher said that we could write about anything. She didn't say it had to be real."

Rosie wished she could make Lucy happy. Lucy hid behind her sarcasm, but Rosie saw her sadness when she talked about her dad. As they walked to their next class, Nathan caught up with them in the hallway. Rosie's cheeks flushed. Nathan smelled like pine trees and she wondered if that was his deodorant or his cologne. The rest of the boys in her class smelled like dirty socks.

Nathan leaned in to their conversation. "What do you think Mr. Barclay's big experiment will be today?"

Nathan loved science as much as Rosie did, and he had helped Rosie with her Plant a Tree project for Earth Day. She hoped they could do a science project together for this year's science fair. Nathan's dark brown hair was perfectly combed to the side and he always wore a button down shirt. He pushed his glasses up his nose repeatedly and usually carried at least three more books to class than he needed. An adorable nerd.

Lucy nudged Rosie's ribs as they walked.

Rosie couldn't look into Nathan's eyes. "I can't wait to see what Mr. Barclay's experiment will be. I loved when he put the rubber balls in liquid nitrogen and they exploded all over the floor!"

"I know. That was so awesome!" Nathan said.

15

Rosie and Nathan babbled away while Lucy stared at her phone unable to contribute to the conversation. Lucy interrupted with a burst. "Rosie has something living in her forest, and we're going to go see it this weekend. Want to come, Nathan?"

Rosie almost barfed in the hallway.

Nathan stopped walking and stared at Lucy. And then he turned to Rosie. "What is she talking about?"

Rosie flashed Lucy her look of 'I'm-going-to-kill-you-later' and then turned to Nathan. "Well . . . we moved out to the country recently, and I'm hearing songs from our forest." The words flowing out of her mouth sounded insane, and Nathan would probably laugh in her face and run the other way.

But he didn't. He wanted to know more. "Whoa. What did you hear?" Nathan's face lit up with curiosity.

Rosie pulled out her journal and told Nathan about the missing Dum-Dums and the strange sounds. Nathan read the song lyrics at least five times.

"Can I come check it out? Are you sure that's okay with you, Rosie?" Nathan looked at the floor and kicked at a crack in the tile.

"Absolutely." Rosie overflowed with joy. Now she would be forced to thank Lucy later.

They walked into science class and Mr. Barclay was in the front of the room setting up tables for his experiment. Everyone gathered around.

"Listen up, class. If you try this at home, get your parents permission and a good mop." Mr. Barclay held a hand up to his mouth and whispered, "The janitor is going to be so mad at me."

Everyone scooted closer and leaned forward with excitement. Potentially dangerous AND messy!

"I'm going to show you the chemical reaction first, and then we'll discuss it. Everyone step back five steps, or maybe six, and Lucy, can you grab the mop?" He pointed to the corner.

Only a few supplies sat on the table in front of Mr. Barclay. The first thing he pulled out from under the table was a bottle of Coke. He asked someone in the front of the classroom to touch the bottle. "What does it feel like, Tracy?"

"Cold."

"Okay, class, we're going to do our first reaction with a *cold* bottle of Coke."

Next, he pulled out a packet of . . . candy? Mentos chewy mints. He dropped two Mentos into the cold Coke bottle. Within seconds, an eruption of foam and Coke overflowed the bottle.

Excitement rippled through the classroom with instant chatter. Mr. Barclay then pulled out another bottle of Coke.

He asked Robert to touch the bottle. He responded, "It's warm."

Mr. Barclay dropped two Mentos into the warm Coke, and a stronger stream of foam shot out of the bottle and flowed onto the table and floor.

"Okay, Rosie, what did you notice between the two reactions?"

She knew instantly. "The warm Coke gave a bigger reaction than the cold Coke."

"Very good observation. Now, I have two more to show you."

He pulled out Diet Coke. He put two Mentos into a cold Diet Coke and it exploded all over Tina and Brittany in the front row.

"So awesome," Nathan mumbled.

Mr. Barclay then pulled out a warm Diet Coke and FIVE Mentos. "Stand back, kids." He put on his safety goggles and held it out at arms length.

Rosie stood on her tiptoes and held her hands to her face as Mr. Barclay dropped in the candy.

WHOOSH!!

It shot to the ceiling and splattered off the fan. Massive spray of Diet Coke showered the room, and the class erupted in a round of whoops and cheers.

The classroom was a sticky, Coke-filled disaster zone. Everyone mopped and wiped up the tables while Lucy stood watch in the hallway for the janitor.

When the mess was gone, everyone sat in their seats so Mr. Barclay could explain the chemical reaction to the class. Rosie sat on the edge of her seat and absorbed every word. Mentos candy was covered with tiny holes and when it was dissolved in liquid, it broke the surface tension. Water particles ruptured and allowed carbon dioxide to escape forming the bubbles and foam. The more candy added, the bigger the explosion. Warmer soda caused the particles to expand for a bigger reaction, and the sweeteners in diet products triggered more force.

Rosie was blown away and Nathan glanced at her with an equally amazed grin as he pushed up his glasses. Lucy filed her nails and did not seem impressed, while Rosie couldn't wait to put Mentos and Diet Coke on her mom's grocery list. After class, Rosie told Mr. Barclay how much she enjoyed him as a teacher and looked forward to the Science Fair next year.

School was out for the summer. Rosie sat on the bus and decided she would miss the teachers and the dusty bus, but she couldn't wait for her summer to begin.

She had a mystery to solve.

Chapter 3

Rosie peered out the kitchen window as they ate dinner that night. She squinted to see the forest in the dim light and strained her ears to hear any sounds.

"Rosie Posey, did you hear me?" Her dad nudged her arm. She had no clue what he was talking about.

"Um – no." Rosie pushed around her peas.

"I thought we could go hiking in the morning and Mom could pack a picnic."

Rosie leapt out of her seat and screamed with joy. "Yes!" Rosie danced around the kitchen while her parents laughed. She immediately started planning their hike.

"Can I bring Baby Annabelle?" Maddy's baby sat at the dinner table on a miniature high chair while Maddy pretended to spoon feed her peas.

Rosie deflated from elation to dread. "Maddy's going, too?"

Looks of warning flashed from both parents simultaneously. Rosie knew the answer.

Maddy would slow them down and demand constant attention. "Can I bring a couple friends along?"

Her mom perked up. "Of course you can! Who are you asking?"

"Lucy and, um, and . . . Nathan."

Her mom's cheesy grin turned Rosie's stomach. Her mom said, "How nice. This will be fun."

Rosie bounced up to her room with new motivation. Her head swam with possibilities of what they might see in the forest. She kept her windows open all night listening for further songs or instructions from the trees. She barely slept as she plotted the hike and replayed the song in her head:

Fear not young lass, we are your friends
We seek an ally to help defend.

Our world's in danger from an evil foe
Please help us soon, or we're forced to go.

She examined every line. *We are your friends, Our worlds in danger, Help us soon.* Rosie needed to help the mysterious people or something terrible would happen. Rosie lost all worry and calm washed over her body. She NEEDED to know more.

At sunrise, Rosie leaped out of bed and got dressed. She pulled her hair into her usual ponytail and wore her comfiest hiking shorts and T-shirt. She also used a touch of *Pink Pout* lip gloss. Nugget bounced at her heels, seemingly aware of an upcoming adventure.

In the kitchen, Mom prepared picnic lunches with sandwiches, fruit, and juice boxes topped off with brownies for dessert. Rosie crept her fingers along the countertop to sneak a snack, but her mom swatted her hand.

"No brownies for breakfast, Miss Chocolate Monster." Her mom looked up and smiled. "Are you wearing lip gloss?"

Of course, her mom would notice!

Rosie touched her lips. "Oh, this? It's the only thing I could find with sunscreen for my lips." Little white lie.

Her mom knew better. "Tell me about Nathan. I don't know him very well."

Stupid lip gloss gave her away. "He's a friend from science class that is super smart. I hope we can team up for the Science Fair."

"Well, then I can't wait to meet him. How's Lucy been?" Her mom's face changed to concern.

"I think she's alright. She doesn't see her dad much." Rosie stared at her hiking boot. "And she's not a fan of Judy."

Her mom wiped the counter for the third time since finishing the picnic preparations. She grabbed the Clorox spray and scrubbed some more. Her mom cleaned with fury and even more vigorously when she worried. "Lucy's Dad is going through a rough time right now, and I'm afraid he's lost without Candice. Lucy needs all the love and friendship you can give her." Rosie's mom kissed her forehead. Her mom and Candice had been great friends, and Rosie knew her mom also hurt after Candice died. Rosie hugged her mom tight.

Rosie's dad sat at his computer in the den. He wore hiking shorts and a faded T-shirt that was markedly different than his usual crisp white shirt and tie that he wore to the bank. "Hey, Rosie Posey, are you ready to explore the timber today?"

"I can't wait! I want to know where the trails go and see what the creek is like."

"The previous owners were an older couple that rarely stepped foot in the backyard. So I'm curious myself." He organized papers on his desk.

Rosie quickly maneuvered to assure her summer was spent exploring. "Dad, do you think that Lucy, Nathan, and I could hike on our own this summer? We'll listen to all of your rules today, and stick to the trails." Rosie was afraid of his answer. If he said no, their whole mission would be ruined.

"I think that would be okay as long as it looks safe today, but you must always remember to tell us when you're going." He put his arm on Rosie's shoulder and they headed to the kitchen.

"What do you think is out there, Dad?" Rosie knew her question carried a bigger meaning.

He responded, "A magical world to explore."

Maddy rolled into the room with her baby stroller and her doll donned a pink dress and hat. Maddy's curls pointed in every direction and she wore purple pants with a green polka-dot shirt. Her doll had more fashion sense.

21

She ran up to Rosie. "Rosie, can I walk with you guys? I want to show Baby Annabelle to Lucy and show her how I can change her diaper."

Rosie cringed inside. "My friend, Nathan, will be here too, and we walk faster than you and Baby… whoever. Can't you stick with Mom and Dad?"

Maddy scowled.

"Her name is Baby Annabelle, and I can keep up!"

Mom intervened with a handful of Dum-Dums – Maddy's kryptonite. "Maddy, can Baby Annabelle hold the suckers today?"

Maddy's eyes grew as she reached for the candy. She licked her lips and practically drooled on the floor.

Her mom winked at Rosie. "Maddy, I hoped that you and Annabelle would walk with me today. We can look for mushrooms and listen for birds, and I bet Baby Annabelle can help."

It worked like a charm.

Maddy bobbed up and down with her bare feet. "Can we pick flowers, too?"

While Maddy was distracted, Rosie escaped from the room. Lucy and Nathan arrived around nine. Nathan's mom had picked up Lucy on the way over because Judy "was too busy" to drive Lucy. They went up to Rosie's room to discuss their strategy. Rosie passed out journals for Lucy and Nathan to carry, and Rosie had her journal pulled close to draw and make notes. She also put a flashlight in her backpack.

"What will we do if we actually see something out there?" Lucy wrote her name in loopy letters with hearts and stars all over her journal.

Rosie had it planned out. "Since Mom and Dad are going, we can't be too obvious if we see or hear something. We need a code word."

Nathan nodded his head in agreement. "How about the word 'bird?' We'll know that one of us saw something suspicious, but your parents will think we only saw a bird."

"What an perfect idea." Rosie smacked her "pink pout" lips.

Mom, Dad, and Maddy were already in the backyard with Baby Annabelle's stroller, and Nugget ran in circles knowing she was going for a walk. Dad carried the picnic basket while her mom carried blankets. Rosie introduced them to Nathan.

Nathan reached out to shake their hands.

Dad squeezed so hard that Nathan winced. "Good morning, Nathan. You can call us Steve and Muriel."

Mom gave him a gingerly handshake. "I'm so happy you're joining us." She then turned to Rosie with a wink.

Rosie looked away and her cheeks felt warm.

They entered the trailhead to the forest exactly where the treasure chest of Dum-Dums had disappeared. Dad paused and turned to everyone.

"Now kids, I'm not sure where this trail leads or what's out here. Please listen to us and stick to the trail. It's supposed to lead right to the creek, but the property did not come with a map. It's a true adventure!"

Rosie breathed deep and instantly smelled wild flowers, damp moss, and pine trees, and she overflowed with excitement. It took all restraint to keep from sprinting ahead on the trail. At first, everyone hiked quietly and the birds, crickets, and fluttering insects filled Rosie's ears. Maddy drowned out the peace with constant yapping at Baby Annabelle and singing silly songs. Lucy swatted at bugs and examined every step she took with her fashion boots and Nathan looked deep into the trees as he pushed up his glasses. Rosie concentrated on the overhead leaves rustling in the breeze searching for any life.

23

Rosie's dad paused to point out poison ivy, as if Rosie didn't know what it was already, as well as sumac and poison oak. "Things to avoid unless you want oozing, itching skin."

"Disgusting!" Lucy wrapped her arms around her body and pulled up the hood of her sweatshirt.

Nathan looked around curiously and occasionally bent down to smell a wild flower, while Nugget bounded from shrub to tree trunk with tongue hanging out. Suddenly, Nugget froze and her fur stood on end. . .

. . . and Rosie heard it.

Off in the distance, a high-pitched melody floated through the forest. Rosie recognized the rhythm from her bedroom window and she looked around wildly to find the source. Nugget barked.

"Do you hear the *bird*?" Rosie's wide eyes locked on Nathan and Lucy.

They stopped and turned to Rosie. Lucy cocked her head and whispered, "A regular bird or a BIRD?"

Rosie nodded her head vigorously while glancing at her parents, ten steps ahead. Her mom and dad chatted, deep in conversation, and appeared to hear nothing.

Nathan hurried forward while looking up into the trees. "I don't hear anything."

"It's far off in the distance. It sounds like a high-pitched whistle, but if you listen carefully it sounds like a song. It's too far away to make out any words."

Maddy appeared at Rosie's feet. "Baby Annabelle heard it, Rosie."

Rosie's insides flipped and she stared at Maddy. They had used the code word, yet Maddy somehow figured it out.

"Did you hear a bird, Maddy?" Rosie tested her.

"No, silly, its the song. They're back." Maddy pushed the doll stroller toward Mom and Dad while Rosie stood speechless. Nathan and Lucy stared in shock as Maddy's wild curls bounced down the trail.

"That's so freaky! Your sister's possessed." Lucy whispered, as beads of sweat appeared on her forehead.

Rosie noticed her parents slowing down. "We can't let Mom and Dad get suspicious. Keep listening, and let me know if you hear anything."

They hiked for half an hour and the mysterious melodies never returned. Rosie surveyed the best climbing trees on the trail, hoping to return to the treetops on a future hike. She mapped out the trail in her journal to remember every detail. She noticed trash on the edge of the trail and stooped to pick it up. Littering was horrible, but who would be on this trail? As she looked closer, her heart skipped beats.

It was a Dum-Dum candy wrapper, and possibly the Dum-Dums stolen from their treasure hunt.

And then the sounds returned.

A clear melody echoed off the trees much louder than previous. Nathan and Lucy simultaneously shouted, "Bird! Bird!" and Nugget barked while running in circles. There was such a ruckus, that Mom and Dad turned around, surprised by the outburst.

Rosie thought fast. "I saw a woodpecker with a dark red head, but it flew away when we yelled."

Nathan smiled and Lucy nodded her head in agreement. Rosie's parents looked at each other and laughed. "That's one heck of a woodpecker." They turned back to the trail.

"Chill out, you guys!" Rosie put her finger to her lips. "I heard it, too. We're getting closer. Could you make out any words?"

Lucy glanced nervously up into the treetops as if wild monkeys would jump onto her head from above. "Not really. The words were blobbed together."

Nathan thought like a scientist and then adjusted the top button of his shirt. "I couldn't make it out, but we must be getting closer to the source. Let's keep moving."

They watched every moving leaf or bush and strained their ears to cut through the normal forest sounds. A clearing in the woods appeared ahead, and Rosie's dad turned back with a thumbs-up.

"We're true explorers, kids. We found the creek."

If only her dad knew the half of it. Rosie sensed the adventure was only beginning. She raced forward to see the creek.

Mom opened the picnic basket on a patch of grass by the water, and started to lie out a spread of food that could feed an army. Rosie stood in the clearing and absorbed her surroundings. A shallow creek rippled with crystal clear water and Rosie couldn't wait to skip rocks on the water's surface. The clearing was tucked inside a protective alcove formed by large boulders and smaller rocks to create a quaint hiding spot in the woods.

A massive bluff lined the clearing along one side and huge boulders lined the base of the rock wall. Across the creek, the dense forest continued without a visible trail. Rosie knew her dad would never allow travel in that direction. Rosie sketched a picture of the clearing in her journal to remember every detail about the bluff and creek.

Maddy played with Baby Annabelle on the rocks by the bluff, while Nathan, Lucy, and Rosie dipped their toes into the cool creek. Rosie daydreamed about swimming and fishing, and hoped they could one day build a raft out of sticks and pretend they were lost at sea. So much potential.

"Where did the song go?" Lucy asked and swirled her pink painted toes in the water.

Rosie scanned the area for clues. "I don't know."

Maddy gathered rocks and piled stones into her stroller. She was a collector. Her closet was filled with treasures from every trip their family had taken. Typically, children want vacation souvenirs like stuffed animals or key chains, but Maddy brought home clamshells, rocks, or a cupful of sand. She swiped the tiny bottles of lotion and shampoo from hotel bathrooms and lined a shelf on her closet with her collections. She currently appeared to be adding more rocks.

Mom hollered that the picnic was ready and sandwiches, juice, fresh fruit and brownies were spread out on the blanket. Rosie's stomach gurgled as they headed for food, but Maddy was slow to leave her baby stroller and rock treasure.

Mom squirted each of their hands with tons of anti-bacterial hand sanitizer before she would hand out a plate. The food tasted perfect, but Rosie knew that the atmosphere of eating in the forest and soaking up the sunshine added to the perfection. Nathan and Lucy devoured everything in sight.

Mom let everyone have two brownies each, and Rosie figured Maddy would be on a major sugar-high after her previous Dum-Dums. Her mom couldn't ignore Maddy's green tongue.

After eating, Maddy immediately returned with Baby Annabelle to the rocks, while Nathan, Lucy, and Rosie wandered to the edge of the clearing with Nugget sniffing constantly at their feet. Rosie's parents stretched out on the blanket, and it was nice to see them relaxing with their arms crossed behind their heads chatting away. They were far enough away and next to the babbling creek that Rosie knew they couldn't hear them talk.

"What do we do now?" Lucy asked.

"I think we search for anything weird. The sounds were louder as we got closer to the clearing, and the trail ends at the creek." Rosie looked toward the tree line again

27

while Nathan headed for the bluff. Lucy checked for cell phone reception.

Then the song returned.

> *Grymballia needs you, please come soon.*
> *Search the stones to enter the ruin.*
> *You'll need the key to go beyond*
> *Grymballia crumbles if you wait too long.*

Nathan and Lucy ran to Rosie, but nobody said a word. Nugget emitted a low-pitched growl, and Rosie scratched Nugget's ears for comfort. Her parents still talked on the blanket making it obvious that they didn't hear the song. Maddy sat on the rocks looking intently into her baby carriage talking and laughing.

"Could you hear the words?" Rosie immediately grabbed for her journal. "Write it down, quick!"

They worked together and wrote down as much as they could remember of the song. They had missed a couple words, but then the song returned. Rosie looked around frantically trying to figure out where the sound came from. She scanned the clearing and the bluff as she listened.

> *Grymballia needs you, please come soon.*
> *Search the stones to enter the ruin.*
> *You'll need the key to go beyond*
> *Grymballia crumbles if you wait too long.*

They corrected their missing words, and they felt confident that the entire song was copied onto Rosie's journal. The mysterious world had a name – Grymballia. Rosie tingled with excitement. Was this really happening? The remote thought of another world hidden in the forest was becoming a reality.

The song came from the bluff.

Nathan pushed up his glasses. "It seemed to come from the rocks by Maddy. Do you think she heard it? She's not acting any different."

Rosie watched Maddy and he was right. At least she was not running to Mom and Dad about anything strange.

Rosie walked toward the bluff. "Let's see if she brings it up."

Maddy sat by her stroller and sucked on another Dum-Dum. There were three wrappers already crumpled on the ground. Rosie picked them up so there was no litter.

"What's up, Maddy?" Rosie asked. "Collecting rocks?"

"We're eating Dum-Dums." Her tongue was now a shade of purple.

Maddy continued singing and talking to Annabelle. Rosie strolled closer to the bluff.

Nathan said, "The song says 'search the stones to enter the ruin,' and the music was loudest over here. Do you think an entrance could be near by?"

"You mean an entrance into another world? You guys, this is ridiculous. This can't be real. There are no other worlds in the forest." Lucy gripped her fists and put them on her hips, but her eyebrows furrowed with a hint of fear.

Rosie knew it was real. "We have to take it seriously, Lucy. What if someone is hurt? They obviously reached out to me for a reason. They need help."

Rosie examined her journal and the words to the song. It said to enter the 'ruin' through the stones, but they would need a key to go beyond. Rosie wondered if there was some kind of door between them and Grymballia.

Nathan agreed. "I have no idea what a ruin looks like, but I think we're on the right track."

Rosie's stomach dropped as she noticed her parents packing up the blanket and picnic basket. They couldn't leave yet! Just then, her parents called them for the return hike. Rosie kicked the rocks as she shuffled back to the blanket.

"Dad, do you think we can come back and explore on our own? We promise to stay on the trail and to not go across the creek." Rosie asked with gritted teeth and a plastered smile and all the sweetness she contained. Hopefully having Nathan and Lucy standing next to her with pleading eyes would also help.

"The trail seems safe enough. Of course, whenever you kids go hiking, I insist you let one of us know, and no swimming without us present. I don't want an accident around the water." He aimed a serious look not only at Rosie, but also at Lucy and Nathan. He made his point clear.

Rosie bounced on her heels. "Thank you, thank you, thank you, Dad." She kissed his cheek. She had permission to explore the bluff in more detail.

"Let's start heading back. It's already mid-afternoon and we have quite a hike ahead of us." Dad gathered up plates and empty juice boxes and put them in a trash bag that Mom had packed.

The songs did not return before they hiked home. Rosie wanted to talk more about the song's words, but her parents were too close. Rosie, Nathan, and Lucy planned to meet up again to hike again soon. Rosie felt it couldn't be soon enough.

When the trail opened into their backyard, Maddy walked straight into the house with Baby Annabelle. She had pushed her jogger the whole way home and did not even whine about it once. She was probably stashing her new rocks in her closet of worldly collections.

Mom drove Lucy and Nathan home. They planned to meet again in two days. Summer break began with a mystery to unravel. Grymballia had sent a cry for help, and it was the adventure Rosie dreamed about.

Chapter 4

That night, Rosie stretched out on her bed and stared at her journal. She examined the drawings of the forest clearing she had sketched and the songs from Grymballia.

Grymballia needs you, please come soon.
Search the stones to enter the ruin.
You'll need the key to go beyond
Grymballia crumbles if you wait too long.

Searching the stones at the bluff seemed an impossible task, especially when they had no clue what they were looking for. *A key to go beyond...* what could that be? Rosie assumed it was not a standard metal door key, but they couldn't go any further until they found it.

Rosie hoped they were moving fast enough, because *Grymballia crumbles if you wait too long.*

Down the hall, Maddy giggled in her bedroom. She had been singing and laughing in her room with her door closed since they returned from hiking. Rosie peeked down the hallway and Nugget stood with tail wagging and with her nose at Maddy's door. Rosie crept down the hallway and opened the door a crack to see Maddy sitting in front of the closet. Five Dum-Dum wrappers scattered the floor in front of her, and Maddy giggled at something in the closet.

Rosie strained to see more, but she could not see anything but toys and baby dolls around a table. Maddy had no clue that Rosie watched her play, and Rosie had a hint of jealousy that Maddy laughed and played without any worries. Maddy didn't know anything about an unknown world hiding in the forest that needed to be saved.

Rosie went back to her room and tried to sleep. Her mind raced with dreams that Grymballia was a beautiful world filled with sunshine and fairies, but also worried that Grymballia could be scary and filled with darkness. Was she doing the right thing? If another world existed, were monsters also real? She didn't sleep well.

The next day, Rosie searched the barn for tools needed to dig at the bluff. She found a small spade, trowels, and grabbed a flashlight. She packed her backpack and threw in her journal for reference. She laid out clothes for the next day's hike and planned to wear long pants to crawl around on the rocks, while also sporting her cute blue T-shirt with endangered owls. She didn't want to sacrifice all fashion sense for the mission.

Rosie desperately wanted a cell phone. Lucy already had one for the last two years and used it all the time. Rosie's parents said she could have one in Jr. High. UGH. She wanted to talk to Lucy and Nathan to pick their brains about the songs, and she was stumped about the key. Rosie wandered around the back yard hoping to hear more music, but she heard only birds and insects. She returned to the house.

Mom was crouched on the kitchen floor scrubbing the grout between the tiles. She was determined to make every last inch of the grout pristine white again.

"Hey, Mom. What's for supper?" Rosie hoped to distract her mom from her misery.

Her mom looked up with sweat on her brow. "Hey Rosie. We're having roast and potatoes tonight."

Not Rosie's favorite. She squatted next to her mom and grabbed a sponge to scrub while they talked. "I was outside getting things ready for our hike tomorrow. Are you picking up Lucy in the morning?" Lucy's dad worked long hours and Judy refused to be Lucy's chauffeur. Nathan's mom was bringing him over.

"I told David that I would pick Lucy up around eight, and he said that he could get her around five tomorrow afternoon. Does that give you guys enough play time?" She wiped her face and dirt smeared across Mom's chin.

Rosie laughed inside because "play time" sounded like hop scotch and hula-hoops when they were searching for a secret world.

"That sounds great." Rosie had only been scrubbing the floor for five minutes and her elbow hurt. "Mom, why don't we tear out this floor as your Christmas present?"

Her mom nodded with a chuckle. "I'm going to take a break for a bit." Her mom rested back on her heels and drank ice tea. "Tomorrow, I might take Maddy to the park to keep her from following you. What are you planning to do in the forest?"

Maddy wasn't tagging along! Rosie wanted to jump for joy. "Thanks, Mom. We're going to hang out by the bluff and will probably be there most of the day. Can we pack ourselves a lunch?" I asked.

"As long as you guys stick together and abide by all of our rules. No swimming and stay on the trails. I've got turkey and avocado for sandwiches if you want to make them up."

Her mom had taught Rosie how to make the perfect turkey-avocado sandwich with a touch of mayonnaise AND mustard with a hint of oregano. Rosie whipped up the sandwiches so that they did not waste a minute leaving in the morning.

That afternoon, a small storm blew through and it cooled off. Rosie loved the smell after a rain - a mixture of dirt and earthworms. She opened her windows to the crisp, fresh air before she went to bed. The stars shined through the treetops signifying clear, sunny skies for tomorrow. The voices did not return, but Rosie rested to the chorus of chirping frogs after the rain.

The next morning, Rosie rode with her Mom to pick up Lucy and couldn't wait to get on the trail. Maddy sat in her car seat combing her doll's hair. Her mom looked into her rearview mirror at Maddy and Rosie knew a battle was brewing.

"Maddy, you and I are going to let the big kids go hiking today, and we can take your babies to the park."

Rosie wanted to cover her ears to silence the protests – but there were none.

"Can we stay home, Mommy? I want to play in my room and have a tea party with my friends."

Shock. Disbelief. Delight.

Her mom was also stunned. "Are you sure?"

"Yep. All my friends are coming to a tea party today. But can I have more Dum-Dums?"

They pulled up to Lucy's house and her dad's car was in the driveway. Rosie hadn't seen him for quite a while, so they all went up to the house. David opened the door and Rosie was shocked by his appearance. He looked so much older and he was pale with scruffy hair on his face. Did he even shower? Rosie sniffed the air.

"David! It's so good to see you." Mom put her hand on his arm and her eyes were full of concern.

He met Mom's gaze, and his eyes instantly welled up with tears. Rosie snuck past him to find Lucy. Mom and David hugged and Maddy pet their dog, Zeus. Rosie suddenly noticed that the cloud of smoke was missing over the kitchen table. Where was Judy?

Lucy was in her room beaming from ear to ear. She ran up to Rosie and tackled her in a hug.

"I'm so excited for our hike," she exclaimed. "Although I don't think we'll find anything."

"Lucy, what's going on around here? Is everything all right?" Rosie knew it wasn't.

"The witch is gone. Dad asked Judy to leave." Lucy could not stop smiling.

"Oh my gosh, Lucy, that's awesome! How come your dad's not happy?" Rosie couldn't pretend to understand.

Lucy's smile faded. "He's so sad, Rosie. He misses Mom so much. He told Judy that he doesn't think he ever loved her, and that she was no good for us. She stormed out in a cloud of smoke."

"Is he going to be okay?"

"I hope so. I love him so much. He's all I have left." She managed a trembling smile.

"You've got me! I'll always be here for you. Maybe we could live together in Grymballia?" With that thought, they bounced out of the room.

Downstairs, David and her mom were in a deep discussion in the kitchen. They ended with a hug, and it appeared that David looked better already.

"Ready to go, girls?"

"Yep!" Lucy said.

"How're you doing, Lucy?" Her mom patted Lucy's head as she climbed into the car.

"I'm good. I'm really, really good actually." They all understood what she meant.

"Your dad's going to join us for dinner tonight. So you girls have all day and all evening to play together. Sound okay to you?"

Rosie and Lucy squealed like three-year-olds and bounced in their seats. They whispered the whole way home discussing their day, and Lucy showed Rosie the supplies in her backpack. Ten minutes later Nathan's mom pulled up and he jumped out of the car with a backpack bursting at the seams. He wore another button down shirt of a different color, he pushed his glasses up his nose, and he grinned.

Let the search for Grymballia begin!

Chapter 5

They stood at the entrance to the forest and stared straight into the deep green darkness ahead. They did inventory of their backpacks. Nathan had packed a shovel, a pickaxe, and a rope, and Rosie carried the flashlight and tools, as well as her amazing turkey-avocado sandwiches. Lucy carried the Band-aids and hand sanitizer that Rosie's mom insisted they bring.

Rosie's first steps into the forest filled her with anticipation and a sense of urgency. Where was Grymballia? And who needed their help?

They set out at a brisk pace and didn't say much at first. The forest life thrived with the previous day's rain. Ferns appeared refreshed and tree frogs whistled from the wet underbrush.

"Where should we look once we get to the bluff?" Nathan asked. He looked sporty in his dark green khaki pants and hiking boots. He wore cool sunglasses and a Gerrard Art Institute backpack from last year's summer camp. Nathan had no siblings, and his parents put him in every art camp possible to prepare Nathan to be a painter, sculptor, or famous artist someday. Little did they know, Nathan only talked about Chemistry.

Nathan continued, "How about we check the clearing first to see if anything obvious has changed. If not, then we start at one end of the rocks and work our way to the other side of the bluff. What do you think, Rosie?"

"Sounds good to me." Rosie said as they hiked.

"I think that sounds good too, if anyone cares." Lucy said loudly with a grunt. "And personally this seems like a waste of time. It's probably someone playing a prank on us and we're walking into a trap. Who's ever heard of another world hanging out in a forest?" Lucy's famous eye roll.

Nathan stopped. "Then why are you here?"

Lucy stammered and shifted on her feet. "The mall was closed today for construction." Then she rolled her eyes. "And maybe I'm a little curious."

Nathan nodded. "Me too."

After hiking for ten minutes, the clearing was ahead. The creek tumbled over the stones, and the water looked refreshing after their muggy hike. Rosie decided to take a break for drinks and snacks before searching the area. They munched on crackers and Gatorade as they faced the bluff so they could absorb the details of the landscape.

Nathan pointed to an area with squinted eyes. "Look up the bluff on the left side, there seems to be a break in the rocks. Maybe it means something is underneath? That might be a good place to start our search."

Rosie could see what caught Nathan's eye. As she looked up the rock face, a crevice formed a large fissure and at its base there were fewer rocks. Rosie said, "Looks good to me." She turned to Lucy. "What do you think?"

Lucy ate her cheese and crackers and looked surprised by the question. "You're asking me?"

"Of course. You're part of the team."

Lucy grinned and looked down for a second. "Do we just start moving rocks and digging? Because that sounds like a lot of work. How do we know what we're looking for?"

"I'm not sure." Nathan answered. "I hope it will be obvious or we'll receive a sign of some kind."

They got to work and headed for the fissure in the bluff. As they approached, large boulders surrounded the base with smaller stones scattered.

Lucy pushed at a boulder with all of her energy. It didn't budge. "Are you serious? We can't move these."

Rosie tried to shove one of the boulders and her shoulder stabbed with pain. They needed a team effort. All three of them pushed a large boulder and managed to move it about three feet to clear a small path toward the base of the bluff. They maneuvered a second boulder out of the way as the sun beat down and sweat poured.

The access to the bluff had opened up, and now they moved away the smaller stones. Rosie and Lucy rested periodically, but Nathan kept working without a break.

The crevice dove deeper into the ground and they could now see an area hidden below the bluff. They dug faster and suddenly Lucy squealed.

"Hey guys! Come here quick." Lucy held up a Root Beer-flavored Dum-Dum wrapper.

They passed it around.

"Maybe it's one of Maddy's from the other day? She ate about ten, didn't she?" Nathan said.

"Maybe," I said. "But we're fifty stones deep into the bluff?"

Lucy said, "The wind?"

"Lucy, remember that all of our Dum-Dums went missing the other day during our treasure hunt."

They stood in silence.

A chill raced down Rosie's spine and she knew the wrappers meant that they were getting close. A tinge of fear crept inside as goose bumps bopped up on her arms, but deep down she couldn't wait to see what they found next.

They dug deeper and every muscle in Rosie's back and legs ached. With every stone that they cleared away, the crevice in the rocks dove deeper and deeper into the earth. Rosie found a cherry Dum-Dum wrapper at the same time that Nathan found lemon-lime.

"It has to mean something. Someone stole them." Rosie rubbed her temples.

Nathan pointed. "I think this crevice is leading to a cave. Look how it dives under the edge of the bluff."

Lucy grabbed another stone. "Let's get in there."

40

Under the bluff was a dark space hidden by the stones. A small cave big enough for one person to slip inside.

"Who's going in first?" Lucy said as she took a step backward.

"I'll go," said Nathan. He grabbed his flashlight from his pocket.

"Wait a minute," Rosie said as she stepped forward. "I want to go in first."

"You two are nuts." Lucy looked back and forth. "You're arguing who gets to crawl into a pitch-black hole when you have no clue what's inside?"

She had a point.

Rosie gave in. "Nathan, you go in and tell us what's inside."

"Are you sure, Rosie?"

Rosie nodded but leaned forward so she could see every step Nathan took.

Rosie stood at the entrance to the cave as Nathan shined his flashlight ahead of him. He eased into the opening and immediately called out.

"You will NOT believe this! There are at least twenty Dum-Dum wrappers in here!" His voice filled with pure disbelief.

"What! Are you serious? What else do you see?" Rosie yelled into the darkness but could not see Nathan. She added, "And be careful."

It was quiet for what felt like fifty minutes but was probably only two.

Nathan finally called back. "It's just an empty cave. I don't see anything else."

Rosie wanted to see for herself. "Can I look around?"

"I'll be right out." Nathan slowly crawled out of the cave's mouth covered in mud.

41

Rosie grabbed her flashlight and wasted no time easing into the opening. She could hear her heart thump in the darkness. The pile of wrappers was scattered in the corner of the cave. Rosie picked them up and put them in her pocket because she couldn't imagine leaving that much litter lying around. The cave smelled damp and water trickled down the rock from yesterday's rain. The cave walls were the smooth rock face of the bluff. Rosie shined her light on every last corner of the cave before she crawled out.

"Lucy, do you want to take a look?"

"Will you promise to stand right here?" Lucy locked eyes with Rosie.

"I won't move. It's pretty cool in there."

Lucy crawled inside. "This might give me nightmares. Oh, my gosh, it's really freaky in here." She yelled from inside.

Lucy didn't stay inside the cave long. Exhausted and hungry, they walked over the creek to have lunch.

"We're definitely in the right spot." Rosie thought out loud. "Finding a cave with a pile of sucker wrappers can't be a coincidence."

"I think it's the right spot, but what are we missing?" Nathan said as he devoured his turkey sandwich.

Rosie pulled out her journal and read:

Grymballia needs you, please come soon.
Search the stones to enter the ruin.
You'll need the key to go beyond
Grymballia crumbles if you wait too long.

"We need the key!" Rosie almost spit out her turkey and avocado. They had discovered the *ruin*, but *needed the key to go beyond.*

42

"But what kind of key, and how exactly do we find it?" Lucy questioned.

"I have no idea." Rosie's shoulder's slumped.

"At least we found the first piece of the puzzle. We just need the key to move on." Nathan was so calm.

Rosie looked up at the sun that was dipping slowly in the sky. "It's getting late, and I think we've done all we can for today." Rosie didn't want to quit, but she had to get home on time so her parents allowed her to hike again. "Can you guys come out again tomorrow?"

"I'll have to ask my mom. She wants me to paint a portrait of Biff." Nathan squeezed his fists.

Lucy almost choked on her turkey sandwich. "You have to paint your dog?"

Nathan dropped his head and his ears turned red. "My parents want me to sharpen my skills with acrylics."

Rosie dropped her voice. "Do you want to be an artist, Nathan?"

Nathan looked at Rosie with sad eyes. "Nope."

She didn't know what to say.

They packed up lunch and their backpacks and washed off the mud by splashing water from the creek. With one last glance toward the bluff, they headed toward the trail leaving continued mystery and a key to discover.

Chapter 6

By the time they arrived to her back yard, her arms and legs ached from all the heavy lifting and she tasted the dirt on her face. Her dad was on the back patio.

"Look at the forest explorers!" He beamed. "Did you roll in the dirt?

Rosie wiped her face with her sleeve. "We've been on the trail and in the clearing. We climbed around in the rocks and had a picnic."

"Sounds like a great day, but you better not step inside the house looking like that. Your mother will attack like Bigfoot if you get mud on her clean floor."

Rosie forgot about the clean floors. "We'll come in the garage door. Can you call Nathan's mom to come get him? Please, please."

"I suppose, Rosie Posey. Anything for the rugged adventurer." Her dad turned and went back into the house.

Nathan grinned at Rosie and she dreaded his next words.

"Rosie Posey?" He held his belly and laughed so hard he almost knocked his glasses off.

Rosie's cheeks warmed and she vowed to strangle her dad next time she saw him. So embarrassing.

"I think it's great," Nathan continued. "Your family is nice. You're lucky."

"You ARE lucky, Rosie," Lucy added. "I feel like I'm part of your family when I'm at your house. I wish my dad called me a silly nickname. Right now, I barely exist."

Rosie wanted to lighten the mood. "My dad calls Maddy his Fart Blossom. I guess I could have it worse."

Nathan collapsed into the grass laughing even harder.

Nathan sat on the porch to wait for his mom, while Lucy and Rosie stepped inside the garage door. Rosie's mom came running.

"Don't move!" Her mom's dirt sensors had alarmed.

She put her hands on her hips. "Who are you and what did you do with my clean princesses." She threw robes to Rosie and Lucy. "We're having a nice dinner tonight, and Lucy's dad is joining us. Why don't you two go jump in the shower?" Mom smiled and then added, "I also asked David if Lucy could stay overnight tonight, and he agreed."

The girls screamed with delight. Rosie lunged to hug her mother, but her mom jumped backward holding her arms up with the reflexes of a cheetah.

"Oh, no you don't! Hugs after your shower."

After washing to remove layers of the bluff and with tunes blaring in Rosie's room, they chatted about the day.

"If the cave was empty, do you think the key is over by the creek?" Lucy brushed her hair repeatedly. She insisted on one hundred brush strokes every day.

"I have no clue. Maybe we'll hear another song to tell us." Rosie pulled the Dum-Dum wrappers out of her pocket and dumped them into the recycling bin.

Down the hall, Maddy laughed in her bedroom. Rosie realized that Maddy hadn't come down to meet them when they came off the trail. Another giggle. Something was strange.

After whispering to Lucy, they agreed to spy on Maddy. Nugget stood with hair on end outside of Maddy's door.

Rosie pushed the door open a few inches. Maddy sat in front of her closet, singing and laughing. Her tea set was arranged in front of her on a miniature table surrounded by dolls.

Rosie pushed open the door, and Lucy and Nugget followed her into the room.

"What's up, Maddy?"

Maddy leaped up and ran over to Rosie.

"Did you come to join our tea party?"

Rosie looked to Lucy and shrugged. "Sure. What kind of tea party are you having?"

Maddy sat in her tiny chair. "Baby Jilly and Baby Annabelle wanted to meet my new friends."

"Did you get a new doll?" Rosie asked.

"No, silly. It's Giblet and Princess Nilly," Maddy drank her imaginary tea.

Rosie and Lucy looked at each other with a look of confusion and a shrug. To get moved up to the tea party table in Maddy's world was a great honor. These must be important guests.

"Where are Giblet and Princess Nilly?" Rosie asked as she looked around to see only doll's eyes staring straight ahead.

Maddy pointed into her closet. "Right here." Immediately, Nugget growled and barked but stayed trembling at Rosie's feet.

Rosie took another two steps forward and squeezed Lucy's arm. Lucy screamed and her knees almost buckled. Rosie's brain could not process the strange tea party scene in Maddy's closet.

Baby Annabelle sat in her highchair, dressed in a frilly pink dress, with a teacup propped in her hand. Baby Jilly sat in a purple plush chair wearing a cheerleader uniform and pigtails with a teacup between her arms. And there, on the far side of the table, sitting on top of Maddy's Legos were two magnificent creatures.

Lucy ran for the door, but Rosie caught her arm to stop her while she also quieted Nugget.

"Lucy, wait! I think this is who we've been looking for." Rosie knew it had to be related to Grymballia. She sat Lucy down on Maddy's beanbag.

46

Maddy drank her tea and giggled.

"Maddy, can you introduce us to your new friends?" Rosie's voice shook and she lowered to her knees next to the tea party while she stared at the amazing creatures.

"They're lots of fun, Rosie. This is Giblet." Maddy pointed to the small frog-like thing sitting on a red Lego.

Giblet was chubby and only about eight inches tall. His smooth green skin looked slimy, but Rosie didn't want to touch him. Beneath his round belly were skinny legs dangling from the Lego with long yellow toenails. Four horns projected from the top of his head pointing in all different directions. Rosie gasped when she looked into Giblet's eyes. His eyes were deep blue and warm, and they made Rosie feel at ease. His large mouth contained no teeth as he sucked on a Dum-Dum.

He waved. "Hi, Rosie," said Giblet. His miniature voice was low and powerful.

Rosie struggled to speak. "Hi." She had so many questions.

Maddy continued the introductions. "And this is Princess Nilly."

The Princess hovered next to Giblet and was only about six inches tall. Her tiny wings slowed as she lowered herself onto a Lego. Princess Nilly was a gorgeous blend of pinks and reds with a petite, human-like body. Her hair was made of leaves in beautiful fall colors. Rosie recognized maple and oak leaves. Princess Nilly's arms and legs looked like sticks, but they were jointed and moved like human arms. Her eyes glowed a radiant gold – like sunshine. She embodied the entire forest in one tiny Princess.

Rosie was fascinated and grabbed her journal to sketch a picture of Giblet and Princess Nilly.

Princess Nilly spoke. "It's a pleasure to finally meet you and Lucy. My people of Grymballia await your help during this difficult time."

Rosie swallowed hard and found her voice. "Princess Nilly, are you the leader of Grymballia?"

"Yes. My parents are the true king and queen, but Plyrim soldiers have kidnapped them. So I'm in charge now."

Lucy gasped. "Kidnapped! Soldiers?" She looked at Rosie with horror.

"Plyrim is the evil country that has invaded Grymballia. Giblet and I escaped to get your help. We need to save Grymballia."

Rosie stared at Princess Nilly's fluttering wings and Giblet's toothless smile. Maddy sat feeding Baby Annabelle a bottle. Maddy was so calm . . .

"Maddy, when did you meet Giblet and Princess Nilly?" Rosie started to piece it all together.

"When we hiked with mommy and daddy."

Maddy explained that she had heard the songs and knew it was coming from the rocks, and then she SAW Princess Nilly flying around the bluff. Maddy offered the Princess a Dum-Dum, but as soon as she opened the sucker, Giblet ran out from the bluff. Giblet jumped into her stroller and snuggled next to Annabelle (who held all the Dum-Dums.) Giblet hummed as soon as a sweet strawberry Dum-Dum hit his tongue. Maddy invited Giblet and the Princess to ride home in her stroller, and they had been hiding in her room for two days.

Rosie stared at her sister with disbelief. She pulled this off without alerting their parents. Impressive.

"Maddy, you can't tell Mom and Dad about this, okay?" Rosie looked into Maddy's eyes.

"Why not?"

"Mom and Dad might not understand Princess Nilly and Giblet. We wouldn't want Mom and Dad to take your friends away, would we?"

Maddy's eyes grew big, and she looked to Giblet and the Princess. "I won't say anything! I promise."

Giblet hopped off his Lego and headed for Maddy's stash of suckers on the tea party table. He walked with a swagger, and it was then that Rosie noticed his long tail studded with more horns. The last horn on his tail was about three times longer than the rest. Giblet grabbed a sucker, walked back to his Lego, opened it, and popped it into his mouth. He hummed as the sugary juices dripped down his chin.

Rosie looked to the Princess. "Princess Nilly, we found a cave entrance yesterday, but we need the key. Who is singing the songs?"

The princess zipped into the air clapping her hands and hovered by Rosie. "I knew you'd find it! The cave is a hidden chamber allowing access to our world, but only with the key can you enter Grymballia. Our fellow Grymballians sing on the other side." She hovered right in front of Rosie's face, and then she floated effortlessly across the room. The leaves in her hair flowed with her every move, and the pink and red colors blended in a blur as she flew. Beautiful.

Lucy finally sat upright. "So where's the key?"

"The key is right there." Princess Nilly pointed to Giblet who waved his hand.

Lucy and Rosie looked at each other, and even Maddy tilted her head in confusion.

"Giblet doesn't look like a key." Maddy said.

"Giblet is a Fimballian. Only the Fimballian magic can get anyone in and out of Grymballia."

They looked to Giblet sitting on the Lego. He pulled his sucker out of his mouth, and a stream of purple drool dripped onto his chubby belly.

"Magic?" Rosie asked. "What kind of magic?"

"The Fimballians have special powers in their horns. With a tap of their tail, they can produce many forms of magic." The Princess waved her arms for emphasis. "Giblet has great power."

Rosie began to make sense of it all. "So we need to bring Giblet to the cave in order to get into Grymballia?"

"That's correct, Rosie. We must move quickly to save Grymballia from Plyrim."

Just then, Dad yelled from downstairs. "Girls! Supper is ready."

Lucy and Rosie looked to each other. Did the Grymballians just hang out in the closet while they ate pot roast?

Princess Nilly read the concern on Rosie's face.

"Go to your family. We will hide out here in Maddy's room. Have no fear."

Maddy, Lucy, and Rosie closed the door and started down the stairs. She had to drag Nugget away as she sniffed around Giblet and tried to lick the sugary juices that dropped onto his belly. She tugged on her collar to get her to leave. Rosie looked at Maddy and put one finger on her lips. "Remember, sshhhhhh."

Maddy nodded and said, "Shhhhhh."

The Grymballians would be there secret. An amazing secret.

Chapter 7

Downstairs, David stood in the kitchen ringing his hands and looking nervous. He was showered and cleanly shaven and wore a nice pair of beige pants with a button down shirt. It was a big change from his morning look-like-a-bum appearance. Mom handed him a glass of wine, and he chatted to Mom and Dad. As Rosie, Lucy, and Maddy joined the group, David pulled Lucy into a tight hug.

"How was your hike in the woods?" David said. "I hear I get the house to myself tonight since you're staying overnight."

Lucy glowed to receive her dad's attention. "We had a great day. My legs feel like cement from all the walking. You don't care if I say overnight?"

"Of course not. So did you break any nails on your hike? It's not your usual activity if it doesn't include hairspray or fingernail polish."

Lucy laughed and slugged her dad in the arm. "Dad!" Then she glanced at her nails.

"It was so nice of Muriel and Steve to ask us for dinner, it will be a great change from cheeseburgers from Maloney's." David squeezed Lucy's shoulder. "Things are going to change at home. I promise."

An awkward silence followed, but Rosie's mom jumped in. "You're welcome to join us anytime for dinner."

Rosie said, "Or I could join you at Maloney's." Rosie loved their chicken strips.

Rosie's dad walked in from the patio with a platter full of bratwurst. "Let's eat!"

"So we're not having pot roast?" Rosie couldn't hide her excitement.

"Change of menu tonight." Mom busied around the kitchen making everything perfect and even set out her good china for the occasion. Eating brats on china made

51

Rosie giggle inside. Mom made corn-on-the-cob and baked beans, plus her special spinach salad with fresh strawberries and slivered almonds - even Maddy would eat Mom's spinach salad.

David shoveled in food as if he was starving, and Rosie thought that color returned to his cheeks almost instantly. David cleaned his plate and devoured a second bratwurst before Rosie finished one.

Everyone leaned back and rubbed their full bellies, and then Mom pulled out dessert. She rested an Oreo ice cream cake in the middle of the table. David may have drooled a little, and Rosie immediately thought of Giblet drooling on his Dum-Dums. Everybody made room for dessert. David praised Mom for her feast for a solid ten minutes. Mom blushed.

After everyone cleared the dishes, David headed toward the door. "Behave tonight, okay?" He gave Lucy another hug.

"I will. Do you care if I go hiking with Rosie again tomorrow?"

"No problem, just be careful out there. The forest is a whole other world."

Rosie and Lucy smiled at each other – if he only knew.

David thanked Rosie's mom another forty-six times and then left. Rosie had not seen him smile so much in months.

"Mom, can we call Nathan to see if he can come hang out tomorrow?" Rosie couldn't wait to tell him about Giblet and Princess Nilly.

"You can use your father's phone."

Rosie and Lucy grabbed her dad's cell phone and ran into the den for privacy. They dialed Nathan.

"Nathan, it's Rosie. You will not believe what happened." Rosie whispered and made sure the door was closed tight.

"Hey, Rosie! Give me a second." She heard him shuffling around. "I'm closing my door." Pause, pause. "Okay. What's going on?"

"We found the key!" Rosie yelled too loud and glanced to the door.

"What? I just left a couple hours ago. Where did you find it?" His voice rose and he spoke faster.

Rosie babbled. "The key is about eight inches tall, green with a big belly, and has a long tail with horns."

Silence.

Rosie continued. "Maddy found two creatures on our first hike, and they're from Grymballia. Giblet and Princess Nilly have been hiding in her room."

Silence.

"Nathan?"

"Uhhhh...you're joking, right?" Nathan's stammered.

"I'm totally serious. I wouldn't have believed it either if I didn't see them with my own eyes."

He hesitated. "Unbelievable. Are they friendly?"

"Yes, and I can't wait for you to meet them tomorrow. Can you come over?" Rosie hoped he had convinced his parents.

"My mom was not happy about it, because she bought a new book on Vincent Van Gogh that she insists I read as soon as possible." Nathan sighed.

Lucy listened over Rosie's shoulder and laughed. "Tell your mom that exercise and fresh air are also recommended for kids."

Nathan said, "Exactly! She said I could come, but I don't know if she'll let me over again anytime soon."

Rosie jumped for joy. "Hopefully we won't need more days. We are going to find Grymballia tomorrow."

"There's really a place called Grymballia?"

"That's where we're headed."

Nathan agreed to arrive early in the morning. Rosie and Lucy threw on their pajamas and hurried back to Maddy's room. Giblet stood on Maddy's Hello Kitty lunchbox wearing a Barbie dress and tiara wrapped in a feather boa. Maddy blinked her flashlight off and on making Giblet glow like a strobe light.

"MADDY! What are you doing?" Rosie threw her hands to her face.

The fashion show stopped.

"Giblet is a model."

Lucy said, "You are a bigger diva than I am, Giblet!"

"Maddy, I'm sure that Giblet doesn't want to play super model." Rosie unwrapped Giblet's boa.

Giblet stepped off the lunch box. "Actually, it was pretty fun." Giblet did not protest as Rosie removed his Barbie shoes.

Princess Nilly fluttered around the room. "If we make it to Grymballia tomorrow, you must realize the danger. My people are hiding, and Plyrim soldiers patrol Grymballia driving everyone out."

"Why does Plyrim want Grymballia? Don't they have their own – world or something?" Lucy asked.

"Grymballia is lush, green, and beautiful because we use our resources wisely. We preserve our water and grow our own food, and we use sunlight for our energy. Plyrim is filled with blackness." Princess Nilly rested onto Maddy's jewelry box.

Giblet continued, "Plyrim destroyed their world. Their water is polluted and not drinkable, and they cut down all of their trees. There's barely enough food, so they

54

need to move away and start over. They decided to steal Grymballia instead."

"We have to help them, Rosie." Maddy broke the silence, but spoke what they were all thinking.

Rosie knew in her gut that she was meant to face this battle and save Grymballia. Her confidence grew even though she had no plan. "We'll help save Grymballia."

Maddy turned to Rosie with puppy dog eyes. "Can I come too, Rosie?"

Rosie turned away and couldn't face Maddy. She didn't want Maddy to slow them down, whine, or demand to go home after ten minutes of hiking. "I don't know, Maddy. It sounds dangerous." Which was also true.

Maddy cried.

Princess Nilly flew toward Rosie. "We need Maddy. Giblet has developed a friendship and follows Maddy wherever she goes. He wants her to come along."

Giblet grinned while Maddy ran over to Giblet and kissed his tiny head. His blue eyes lit up even brighter and his tail shot out a spray of sparks. The room went dark.

Rosie heard Lucy gasp. The sound of three thumps came out of the darkness and sparks lit up the darkness. The lights went back on.

"Whoops!" Giblet said.

Magic? Giblet could smack his tail and make things happen. Rosie stood in awe.

"You can come, Maddy, but we have to be careful and you'll need to listen to me." Rosie knelt in front of Maddy to look in her eyes. "Okay?"

"Okay, Rosie." Maddy threw her arms around Rosie's neck and Rosie couldn't remember the last time she had hugged her sister.

Rosie and Lucy crawled into bed and chatted about the next day's adventure. Quiet followed as Rosie's mind spun with the possibilities of what Grymballia might bring.

Chapter 8

Nathan arrived bright and early riding his bike with a packed backpack. "I'm came as soon as I could get away," he said, breathless. "I couldn't sleep all night."

Rosie knew the feeling. She pulled Nathan inside and she and Lucy led him to Maddy's room to meet their new friends. They heard singing down the hall and assumed it was Maddy, but when they opened the door, Princess Nilly hovered overhead with her brilliant wings fluttering. Her golden eyes glowed as the soft melodies bounced off the walls.

Nathan watched with wide eyes and mouth gaping open. He could only point and was unable to speak. When the princess spotted us in the doorway, she stopped singing.

"Don't stop!" Lucy pleaded. "That was amazing."

"You're kind, Miss Lucy, but I see that Nathan has arrived."

Nathan looked to Rosie, still speechless.

"Nathan, meet Princess Nilly," Rosie said.

Nathan bowed to the princess and Rosie wished she'd thought of doing that.

Giblet sat on a toy stuffed horse next to Maddy. Baby Jilly snuggled on Maddy's other side with a pacifier in her mouth. Giblet straddled the horse and held the reins as he wore miniature cowboy boots and a hat propped on his horns. Nugget stretched out on the floor next to Maddy and Giblet was oblivious to how strange it all looked.

"Nathan, this is Giblet on the horse," Rosie said, pointing.

Giblet tipped his hat in howdy. "Nice to meet you, partner." Giblet said. "Glad you could join us."

Nathan stared at Giblet. "I – I – don't understand." Nathan rubbed his head and leaned against the wall.

Rosie filled him in on the history of Grymballia and Plyrim. Plyrim soldiers kidnapped the King and Queen of Grymballia to overtake their world because they had destroyed their own environment. They must enter Grymballia using a key, which was a creature of Fimballian descent – Giblet, who also had magical powers in his tail.

It was overwhelming, but Nathan absorbed it all as he pushed up his glasses and nodded in silence.

"How are we going to get the King and Queen back from Plyrim?" Lucy asked.

Princess Nilly rested on the jewelry box. "The rescue will depend on all of you reaching the castle and bringing the King and Queen to safety."

Nathan came alive. "Let's get to Grymballia first, and then we can assess the situation."

"Will the Plyrim soldiers have weapons?" Rosie glanced to Maddy and her stomach flipped. She worried what they were getting into.

The Princess understood. "Our world is different. Our weapons are not mechanical tools that cause harm, but magical powers put us at risk."

"Everyone has magical powers?" Nathan asked.

"Yes, but we are gifted in different ways," Princess Nilly said.

"Would weapons from our world work?" Nathan paced the room.

Lucy gasped. "What do you mean, Nathan?"

Rosie squirmed and didn't like the conversation. "Yeah, what do you mean?"

Nathan put up his hands in defense. "No! I would never carry a real weapon, but we need to figure out how to battle magic."

"Most are protected from harm with charms and shields, your human weapons would have no effect."

"Thank you, Princess. We're going to have to use our brain as our weapon." Nathan picked up his backpack. "We should get going."

Giblet dismounted his horse, removed his cowboy hat, and grabbed a sucker for the road. Maddy put Baby Annabelle in her stroller and made room for Princess Nilly and Giblet to ride inside. They moved through the kitchen quickly to avoid a curious mom from asking too many questions.

They didn't make it too far.

"What's the rush? The woods aren't going anywhere." Rosie's mom paused. "Maddy, you're going?"

Rosie's mom glanced toward her to assure it was approved.

"It's okay Mom. We'll keep a close eye on her. We told her she could join us today," Rosie said, looking at Maddy. "As long as she listened and stayed out of trouble."

"Okay then, I packed your lunches." She handed Rosie a heavy picnic basket. "I threw in some Dum-Dums for the road."

Giblet rustled in the stroller when he heard "Dum-dums," and now his head peeked up over Annabelle's ear. Rosie waved him back down before her mom noticed. That was close!

They entered the forest trail loaded with supplies and Nugget sniffed while leading the way. The crisp morning air felt wonderful on her cheeks and Rosie was charged with excitement. She barely noticed the squirrels or wild flowers because she practically ran down the trail.

The clearing was scattered with rocks, and Maddy noticed immediately.

"What did you guys do? Can I climb on the rocks?" Maddy hurried forward.

"No, Maddy." Rosie said. "We don't have time to play. We need to get Giblet and Princess Nilly to Grymballia."

Nathan and Lucy cleared more stones so they could all fit into the cave. Princess Nilly led the way inside. Maddy carried Giblet like a baby and left her stroller outside. Giblet gripped Maddy's curly blonde pigtail for the ride.

"Who cleaned up my wrappers?" said Giblet. "Your treasure hunt was delicious. I'm sure am going to miss candy when we get to Grymballia."

"I cleaned up your mess. You shouldn't litter, and I don't understand your addiction to Dum-Dums, Giblet," Rosie said. "You eat about twenty a day!"

"In Grymballia, we only eat sweets that we can grow on our own. We have no candy, chocolates, or other artificial sugars," Giblet said. "I'm trying to get my fill before I return."

Princess Nilly laughed a high-pitched giggle. "You'll also notice that Giblet, like most Fimballians, has no teeth. Since they are the keys to leave our world, Fimballians tend to gorge on sweets and often lose their teeth from decay."

Giblet smiled a wide, toothless grin.

Rosie looked around the cave. "Now what, Giblet? How does your key work?"

Giblet jumped down from Maddy and became serious. He sauntered toward the cave's center. "Step back and shield your eyes."

Rosie grabbed Nugget and tried to shield her eyes while assuring that Maddy did the same.

Rosie peeked through curious fingers to see Giblet drag his tail to form lines in the dirt. With his long toenails, he drew a large circle with smaller radiating lines to resemble a sun. He stood in the middle of the circle. Giblet hummed in a low pitch and then started chanting:

> *Land of the Earth, we have come forth,*
> *We bear no harm and promise our worth.*
> *Nature's our friend; we will never neglect*
> *Grymballia we enter and always protect.*

Sparks flew as Giblet smacked his tail on the cave floor. The earth rumbled and the cave shifted.

Maddy grabbed Rosie's leg. "What's happening, Rosie?"

Nugget shook in Rosie's arms. The back wall of the cave glowed a bright blue and transformed into a fluorescent ball.

A portal to Grymballia.

Giblet announced, "Everyone inside." He pointed toward the swirling light.

Nobody moved.

Giblet tapped his foot and crossed his skinny arms over his chubby belly. "You must enter the light to get to Grymballia."

Princess Nilly flew into the light with a flash of pink as a leaf blew from her hair. Then she was gone.

Maddy pointed frantically. "Where did Princess Nilly go?"

Rosie hid her fear from Maddy and acted excited. "She's in Grymballia! Now it's our turn." Rosie planned their approach. She couldn't let Maddy go ahead of her, or she would be too scared. Nathan should enter first, then Lucy with Nugget, followed by Maddy and herself.

Giblet tapped his foot.

Nathan stepped up to the light. "Here it goes." He pushed his hand through the glowing portal and it disappeared. He pulled it back out, and it was perfectly normal. "Whoa!" He grinned at Rosie and then plunged through the blue light while whooping with joy.

Lucy hesitated only briefly before charging through the light with Nugget wrapped in her arms.

Rosie stared at the swirling portal and saw nothing but light. Giblet, the gatekeeper and key, nodded with encouragement. Rosie squatted next to Maddy.

"Okay, Maddy, we'll jump through the light together and see Nathan, Lucy, and Nugget on the other side. Giblet will come right behind us. Are you ready?" Rosie wasn't sure SHE was ready, but she couldn't let Maddy know.

Maddy stepped up to the portal. "Do I get to see where Giblet lives?"

Giblet nodded. "I can't wait to show you my village."

"Will it hurt?" Maddy asked.

Rosie made a hopeful guess. "Of course not."

"You will be safe." Giblet said.

Maddy grabbed Rosie's hand they stood in front of the portal. Giblet gave Rosie a thumbs-up and they stepped into the light.

WHOOSH

Rosie's stomach flipped like she was upside-down on a rollercoaster. Blinding light surrounded them and they were sliding in a cool rush of air. Maddy squeezed her hand and then... THUMP. They landed. Nathan and Lucy reached out a hand to help them up. They were in another cave.

Giblet arrived with another thump, and then smacked his tail three times on the cave floor. The portal's glow disappeared.

They stood in a bigger cave on the Grymballia side, and everyone had made the journey safely. Even Nugget stopped shaking and started to sniff the new terrain.

61

"Are you okay, Maddy?" Rosie asked.

"It was like a water slide," she squealed.

Lucy dusted off her pants and fluffed her hair as she looked around the dim cave. "This is not how I pictured Grymballia." She looked at herself in her phone. "There's no cell service here either."

Princess Nilly floated to the cave's entrance where sunlight poured inside.

She motioned for everyone to follow her. "We must hide our portals inside caves to protect our land from discovery. Let me show you Grymballia."

Rosie shuffled toward the entrance and felt the warm sunshine on her cheeks. As her eyes adjusted to the light, the beauty of Grymballia stole her breath.

"Whoa."

Chapter 9

How could so much color be packed into one landscape? Patches of white fluffy clouds sprinkled a brilliant ocean of blue sky. Rosie could see for miles into the lush Grymballian valley below. Scattered clearings raised question as to what creatures lay ahead.

A large waterfall dropped into a crystal clear lake, and the sunlight captured the mist to form rainbows in the sky. The air smelled sweet from flowers scattering the ground.

Then Rosie heard the music. Not just one beautiful voice, but a choir blended into soft melodies that echoed through the treetops bringing an immediate sense of calm.

"Princess Nilly, I've never seen anything so peaceful," Rosie said.

Nathan and Lucy scanned the area with wide eyes of amazement. Maddy stood next to Giblet.

In the center of the great valley stood a castle. Unlike the stone, square castle of fairy tales, it was a large circular dome building with a great wall extending off in each direction. Gold reflective plates covered its entire surface.

Rosie recognized the castle as the drawing Giblet made in the cave floor. The castle resembled a sun glowing in the center of their world. She sketched the castle and landscape in her journal, but she knew drawing could never live up to the beauty around her.

"Princess Nilly, is that where you live?" Rosie pointed toward the golden palace.

"Yes, but it's not mine. Our castle belongs to the people of Grymballia," the princess said. "We call the castle 'El Sol' which means 'The Sun.' Golden solar panels provide power to all of Grymballia." She dropped her head. "But it is currently overrun by Plyrim soldiers."

63

Lucy's mouth still gaped open. "If we win this thing, I think I'm moving here."

"It's pretty, Giblet." Maddy said. Giblet hung close to her as they chatted. Rosie wondered if he stayed close to Maddy for her protection.

Princess Nilly flew overhead darting in different directions as she surveyed the land. "We must watch for Plyrim soldiers because they are often on patrol."

Nugget sniffed around. Rosie knew they needed a plan now that they had made it to Grymballia. "Let's meet under this tree to stay out of view." She motioned for everyone to follow her under the low hanging branches of a massive tree.

As Rosie scooted underneath, the branches parted on their own to let her pass. She jumped backward with shock. The others hadn't noticed, and Rosie thought she must have imagined it. She stepped under the tree and looked inside. Multiple vine-like branches intertwined within the canopy. Thick knots covered the tree trunk, and Rosie charted the path she would use to climb toward the treetop. She could spend all day in this tree. "What a magnificent tree," she whispered.

"Thank you, Rosie." The deep rumble came from the tree's trunk. Two knots in the bark popped open to reveal eyes looking back at Rosie. Nugget growled and tugged on a vine.

"Nugget, stop!" Rosie pulled on her collar and looked to the knotted eyes. "I'm sorry. Are you really a talking tree?" The words sounded crazy coming from her mouth.

Maddy entered under the tree just as it spoke and held tight onto Rosie's leg. "Rosie, I'm scared."

Rosie patted Maddy's back. "It's okay, Maddy."

"I mean you no harm. Most trees in Grymballia can talk, but only to a worthy listener. You, Rosie, are worthy.

You respect Mother Nature and that makes you our friend."
His vines reached out and Maddy touched his branches
with her fingertips.

Just then, Giblet ran under the canopy.

"Hide us, Franklin! Plyrim soldiers are coming!"
Giblet spoke to the tree. He stood next to Maddy and
watched the horizon. Lucy and Nathan scurried under
Franklin's limbs.

A screeching sound in the distance was getting
closer. Franklin told everyone to gather at his trunk. Rosie
grabbed Maddy and Nugget and held them close, and Lucy
and Nathan rested their heads against Franklin's knotted
trunk. The screeching sound grew from the direction of the
castle. The Princess hovered high in the canopy on lookout.

"They're almost here," she called.

Franklin grew his vine-like branches and wrapped
them around them all for protection. Rosie now knew what
a mummy felt like, but it was the best camouflage. The
screeching sound stopped outside of the tree, and she heard
voices. Rosie smelled them before she could see them as a
horrible stench permeated the vines. Maddy wriggled and
Rosie squeezed her tighter so she would not break free.

Rosie peered through a slit in Franklin's protective
wrap and saw a foul cloud of black dust surrounding two
dark shadows.

A shadow spoke in a gruff voice. "I heard
something. Search the area."

They dared not move. The dust cleared and she
glimpsed the Plyrim soldier. Her heart raced and her hair
stood on end. A massive pile of warts, horns, and ugliness
piled on a mammoth body about seven feet tall. Rosie
hoped that Maddy could not see them for fear she would
scream in horror.

Franklin shifted, and Rosie lost her viewpoint just as the screeching started back up. The sound drifted away into the distance.

When the Princess told Franklin it was safe, he released his wraps and they tumbled out.

One glance to Nathan and Lucy confirmed that they saw the Plyrim soldier too.

"What the heck was that?" Lucy looked to Princess Nilly.

"The Plyrim soldiers patrol to capture any Grymballians roaming free. They probably heard my friends and their welcome songs so they are hunting for them. Grymballians are good at hiding, but if caught, they are thrown into the castle's dungeon."

"Nobody is safe?" Nathan asked.

Giblet stood tall. "Plyrim took over Grymballia. Our only hope is saving the King and Queen. We must make our way to the castle."

"How many soldiers are there?" Nathan asked. "How can we hide from all of them?"

"We must watch close and use our resources to hide because more soldiers guard the castle. But once the King and Queen are rescued, we can regain control of Grymballia. Only their royal power can banish the Plyrim soldiers." The Princess looked directly at Rosie.

Rosie felt a heavy wave fall onto her shoulders, as their task suddenly seemed impossible.

"They stink?" Maddy pinched her nose.

The princess giggled. "Plyrim used up their water resources, so they have not bathed for many moons."

Giblet led the way as they hiked toward the castle. Princess Nilly flew overhead scouting for Plyrim soldiers, even though Rosie knew they would smell them coming.

Rosie elbowed Nathan. "Any ideas? How on earth are we going to free the King and Queen?"

Lucy leaned in. "What are we doing? We can't fight those disgusting things."

Nathan was more confident. "Let's see what's up ahead. I have an idea, but I need more information about this place."

Rosie looked to Lucy and shrugged.

They walked through a forest floor blanketed with green moss, ferns, and flowers. Mushrooms covered with purple, pinks, and polka-dots popped up between rocks. Flowers climbed as tall as Rosie with blooms as big as her head.

Rosie paused at a flower and saw a bumblebee zip inside.

Wait.

That was not a bumblebee!

A winged creature landed on the yellow petal and Rosie stood on a rock to look closer. It had six arms and legs with flowing red hair all over its body. Two antennas projected on top of its head and two large eyes rotated in many directions. The eyes were a piercing blue with long eyelashes. A long snout projected with full, red lips on the end. Suddenly, her stunning sapphire eyes focused on Rosie.

"Hello, Rosie. Are you watching me work?" The red puckered lips formed every word.

Rosie almost tumbled off of her rock. The secrets of Grymballia still surprised her.

"H-h-hello." Rosie said. "What are you doing?"

"Pollinating of course. That's how we keep the flowers of Grymballia thriving. Watch me." She waltzed to the center of the bloom and swayed her long locks of red hair back and forth in the pollen until she was coated.

Rosie leaned forward, fascinated. "Do you carry the pollen to other flowers on your back?"

"That's right!" She smiled her puckered lips, and her blue eyes glowed from the top of her antennae.

Rosie wanted to continue watching, but they must keep moving.
"It was great to meet you. Hope we meet again."

Rosie caught up to the others a short distance ahead. Nugget strolled slowly at her feet and soaked up the strange world while Nathan and Lucy huddled in deep conversation. A spectacular view appeared over the next hillside. A waterfall shot out from a cliff covered with ferns, and the water cascaded into a lake below. They journeyed to the water's edge for a short break and pulled snacks and water from their backpacks to share.

"Rosie, when are we going home?" Maddy asked with a hint of whine.

Rosie sighed. She had been waiting for Maddy to quit and throw a fit. She honestly predicted she would have pitched her tantrum before now.

Princess Nilly flew to the rescue and sat on Maddy's knee. "Miss Maddy, we need your help to save Grymballia. Can you stay strong and help us?" Princess Nilly's little twig arms patted Maddy's leg.

Maddy sighed. "But I want Mommy."

Rosie tapped her foot and glared at Maddy. She should have left her at home. She glanced to the castle and knew time was wasting.

Princess Nilly continued. "Follow me, Maddy. I want to show you how special Grymballia is."

The princess flitted over to the water's edge. She set herself down on a pebble and waited for Maddy. Rosie stepped closer to get a better view.

"Our land and water are valuable." The princess laid her twig-like hand on the water and the lake's bottom started to glow. Fluorescent pink, purple, and orange rocks twinkled underwater. "Maddy, would you like to a take a piece of Grymballia home for your collection?"

Princess Nilly was a genius. Rosie knew that Maddy could not resist more rock treasure to pile in her closet. Maddy peered into the water and squealed with excitement.

"Gemstones! Pretty rocks!" Maddy screamed and pointed. "Rosie, look!" Maddy reached into the water and pulled out a brilliant orange stone, and then reached in three more times to pull out a purple, pink and red. Each precious rock was glowing and radiating with its own unique beauty. She held them close to her chest and ran over to show Rosie.

Rosie couldn't help but grin. "They're pretty, Maddy. Can we save Grymballia now?"

"Can I really take the rocks home?" Maddy looked at the princess.

"Consider them my gift for your help today, Maddy." Princess Nilly nodded.

Maddy gathered her rocks and packed up her things to continue the mission. As they looked at the massive lake, they realized hiking around it would take hours.

"We could swim across?" said Lucy.

"What about Maddy and Nugget?" Rosie asked. "They can't make it that far, and we don't have time to build a raft or anything."

"We'll have to hike around." Nathan said.

Just then, Princess Nilly cupped her hands and her voice echoed into a high-pitched song. She was calling someone. The leaves in the treetops above the waterfall started to rustle. The princess called again. Four shadows rose from the trees and started flying down toward them. The mighty winged beasts landed by the lake.

The Princess called them Blim Birds. They hugged the treetops in Grymballia and were known to be a friendly

69

ally. They stood five feet tall with a wingspan that was twice that height. Their colors included every hue of the rainbow down to the tips of their wings. They had short legs with massive feet that appeared catlike with fur, pads, and hooked claws instead of bird feet. The short body led to a thick neck and a most interesting head.

Rosie wanted to pet the creature to feel its rainbow fur. She quickly sketched the Blim Bird.

Despite being called a bird, there were no bird features present on its head and it seemed more like a cat? It had a small round head with big green eyes. The furry, pointed ears were too large for the Blim Bird's head, and it had whiskers and a small red nose. The mouth was filled with teeth.

Enormous teeth.

The rainbow colors of the winged body continued over the fur of the neck and head. Rosie walked around the Blim Bird and it rubbed its furry head on her shoulder like a cat.

Lucy stood face to face with the creature. "Why are their teeth so big?" Her face couldn't hide her fear.

The Blim Bird spoke and Lucy jumped almost two feet into the air. "We search the forest floor for food to eat, but we'll also eat Plyrim soldiers." The Blim Bird chuckled. "But they taste awful."

"You're a talking rainbow," Maddy said.

"We use the rainbows of the waterfalls for camouflage. Now, let's get you to the castle to save Grymballia. Hop aboard!" The Blim Birds squatted down on one leg to allow each to climb on their back.

"I get to fly?" Maddy said as she bounced up and down.

"We do, but you need to hold on tight." Rosie had never flown in an airplane let alone on a bird. Her stomach flipped as she climbed aboard.

The four Blim birds knelt down and Maddy rode with Rosie, Lucy took Nugget, Nathan rode alone, and Princess Nilly and Giblet rode on the last Blim. They could not ride all the way to the castle or the castle guards would spot them. They took off over the lake and trees.

Rosie looked down at Grymballia below as her legs straddled the furry, yet feathered Blim body. She was surrounded by yellow, blue, orange, and red color and asked if they could fly through a rainbow. The Blim Bird banked toward the cascading waters of the falls where the sun reflected off of the water droplets. Rosie closed her eyes and put her face toward the warm sun and wondered if there was a more perfect place on earth.

Nathan's Blim passed through the rainbow up ahead, and he disappeared! Only a hint of Nathan remained as the Blim's colors disappeared in the rainbow.

"Nathan, watch us!" Rosie yelled.

As Rosie's Blim Bird flew through the rainbow, she could feel the moisture droplets from the waterfall on her face. Maddy giggled and squirmed in Rosie's arms.

"Whoa! You vanished!" Nathan was so excited.

They continued to fly over the treetops. Maddy talked constantly and pointed to every detail in the land below while asking a million questions. Nugget sat on Lucy's lap with her face in the wind. Grymballia was a lush and tropical paradise of dense trees with occasional clearings of houses. Giblet and the Princess Nilly flew next to Rosie, and Giblet pointed to his home village. Maddy almost fell off the Blim Bird trying to see every detail.

"Can we go to your house, Giblet?" Maddy asked.

"Maybe after we save the King and Queen, Miss Maddy," replied Princess Nilly.

Maddy crossed her arms over her chest, but Rosie grabbed her another Dum-Dum. Bribery by candy, but what would she do when the candy ran out? Rosie wondered why they couldn't see anybody below in the villages.

"Giblet, where is everyone in your village?" Rosie asked.

"You can't see everyone from this height since we're so small, and they must hide from the soldiers."

Rosie panicked. "Have they captured any of your family, Giblet?"

"They have my brother, Goblet. He tried to stop the soldiers when they were taking the King and Queen. They threw him into the dungeon with many others." Giblet dropped his head and looked away.

Rosie didn't hesitate. "Then we'll have to save them, too."

Giblet looked at Rosie in shock. "Miss Rosie, you're here to save the King and Queen. You can't risk yourself for my brother." I could see fear in his eyes, but also a glimmer of hope.

"Giblet, your brother is your family. We cannot leave him behind." Rosie had no clue how they were even going to save the King and Queen, but now the stakes were higher with a dungeon full of Grymballian prisoners. "We'll do our best." Rosie saw a tear in his big blue eye.

Nathan flashed Rosie a warning look from his Blim Bird. She knew he was worried, and she nodded her head to say *I don't know how we'll do it yet, but we have to make it work. Trust me.*

Nathan smiled.

The castle neared in the distance and its royal grandeur stole Rosie's breath as the sun beamed off the gold dome. Grymballia stood for everything Rosie believed in, and the castle provided the natural energy for the entire land with its solar panels. At the base of the dome, pillars and archways opened into gardens filled with flowers of every color imaginable.

The Blim Birds set them down about a half mile from the castle to avoid being spotted by the Plyrim soldiers. They landed silently on padded feet, and everyone dismounted with ease.

"That was the best ride ever." Maddy wrapped her arms around her Blim Bird in a colorful hug of thanks.

The Blim gave Maddy a cat-like lick on her cheek. "The pleasure was all mine. Please save Grymballia, will you?"

"We will!" Maddy promised and looked to Rosie.

The Blim Birds flew back to the treetops to be hidden by rainbows. Suddenly the putrid smell of rot surrounded them. The Plyrim soldiers were close.

"What do we do?" Lucy grabbed Rosie's arm.

"Everyone stand close to me. I have a plan." Giblet slapped his tail three times and sparks flew. The soldiers were coming rapidly and their black cloud of filth could be seen over the treetops.

"Don't make a sound," Giblet whispered. "I have made us invisible to the soldiers."

Invisible! Rosie looked at Maddy and Nathan and they were faded and barely visible. The black cloud was getting closer. Was Giblet sure they were invisible? What if Nugget barked?

The cloud of black smoke and the stench arrived and the Plyrim soldiers stood five feet away. Rosie got a closer look and they were horrible. Their face was smashed in with beady, black eyes dark as night. They had long fangs that jutted out of their mouths that were yellow and crusted and decayed. Warts covered their skin and black tufts of hair jutted out of their ears. Two broad nostrils oozed mucus as they breathed with a snort. And they STUNK.

Rosie squeezed Maddy's shoulder and hoped she wouldn't move. Being invisible wasn't something she had time to explain. The soldiers searched around trees and under bushes coming within inches of their invisible huddle. Nugget sniffed the air and withdrew her nose in disgust.

The soldiers turned and moved away and were soon out of site. Giblet had saved the day.

Maddy asked many questions. "Rosie, what was that thing? What do they want? How come they couldn't see us?"

"Maddy, those are bad, smelly soldiers, and we have to hide from them. Giblet made us invisible so they could not see us." Rosie knelt down to Maddy. "But you did great, Maddy. You stayed quiet." Rosie was impressed. It could have been a bad situation if Maddy had freaked out. Rosie scratched Nugget's ears. "You too, Nugget."

"Giblet, you're amazing. Should we just keep the invisibility spell until we get into the castle? We could hide from other solders and find the King and Queen," Rosie said.

"What a great idea." Nathan said. "That way we won't be spotted and can look around."

Giblet got a worried look on his face. "The magic only lasts a short time before it wears off. We might have only an hour."

"Then let's get moving." Nathan commanded.

Princess Nilly stopped the crew. "My friends, there will be more soldiers ahead. The objective is to find the King and Queen, but if it is too dangerous, we will turn back. I will not risk your lives." The princess rested on Lucy's shoulder.

"Princess Nilly, where were the King and Queen last seen in the castle?" Nathan asked.

Giblet answered for her, "They were taken to a room at the top of the castle that is guarded by a Fligarian.

74

Fligarians are nasty, fire-breathing creatures that are hard to defeat."

"What about the rest of the Grymballian prisoners?" Rosie asked. She hoped they could save Goblet, Giblet's brother. "Where is the dungeon?"

Giblet said, "It's below the main level filled with Plyrim guards. The dungeon is fortified with poisonous Ackly vines so escape is difficult."

"This sounds impossible." Lucy looked at Rosie and then Nathan.

"We have to save them!" Maddy ran to Giblet and linked her arm with his.

"We will, Maddy." Rosie agreed with her sister – for once. "We'll get inside the castle and look around while we're still invisible. There's got to be a way we can set them free."

"Once we free the King and Queen, can they help Grymballia?" Rosie asked Princess Nilly.

"The King and Queen of Grymballia have fantastic powers. If they are freed, they can banish the Plyrim soldiers and the Fligarian, and free the imprisoned Grymballians. We simply need to return them to their thrones at the center of the castle." The princess answered.

"Then that's our plan. We free the King and Queen, and get them to their thrones. Once they're free, then the prisoners will be rescued and all of Grymballia will be saved."

Rosie knew it wouldn't be so easy.

Chapter 10

Rosie smelled the sweet aroma of lilacs mixed with roses with the faint stench of Plyrim soldier. The brilliant dome of the Grymballian castle reflected the sun and bathed the flowers with its light. The solar panels hummed as it harvested the sun's energy for all of Grymballia. Rosie held Nugget at her feet and gripped Maddy's hand as they ducked behind a flowered bush close to the castle's main entrance. Plyrim soldiers marched back in forth scanning the forest.

Rosie whispered, "How are we going to get inside?"

Princess Nilly fluttered close to their heads. "This is the main entrance, maybe we could try entering through the water wheels at the back of the castle. It might be another way inside."

"We have to try," Nathan said. "We can't risk our invisibility wearing off trying to get past these soldiers."

"I agree," said Giblet. "I don't know how much time we have left."

As they snuck around the side of the castle, a soldier suddenly appeared and was so close, he almost stepped on Nugget. They froze and hoped the invisibility lasted a little longer. The Plyrim soldier looked around.

Rosie slowly bent over to scoop up Nugget to keep her calm. The smell of the soldiers made Rosie gag and she tucked her nose into her armpit. A little better. The soldier grunted and then moved on. They all breathed a sigh of relief.

"That was close!" Lucy said. "What are we going to do when we're not invisible any more?"

Rosie looked awkwardly to Lucy and Nathan. "Let's just get inside the castle." Rosie glanced to Maddy and couldn't think of what might happen if they were captured. A dungeon was no place for a six year old.

Hurrying around the golden castle, they spotted the river ahead. The water flowed toward the castle and struck a monstrous, wooden waterwheel churning to pull the water inside the castle. The wheel produced electrical energy from the water. No doors or other entries were around, only the waterwheel could get them inside.

Nathan examined the wheel in silence. "It will work. We'll have to ride on one of the wooden slats over top of the wheel to be dropped off inside the castle. We're going to get wet."

Lucy gasped as she fluffed her hair. "Are you sure there's not another way?"

"I can fit over the wheel," said Princess Nilly. "Let me check for soldiers on the other side." She soared over top of the waterwheel and soon came back.

"All clear," she said.

After much discussion, Maddy rode with Nathan so he could hold on better if it was too slippery. Rosie would ride with Lucy while the princess flew over the top. Giblet and Nugget found a wedge opening between the wall and wheel to squeeze through.

Rosie and Lucy went first and walked up to the wheel. The shallow creek rushed by and the loud splashing made it hard to hear each other. The wheel moved faster than Rosie expected. She looked to Lucy and gave a thumbs-up.

"Girl power!" Lucy smiled.

As the waterwheel turned, they held hands and waited for the next wooden platform to come up from the water. They leaped forward and landed together on the board. The loud water splashed their faces and the fast churning wheel made Rosie panic.

"It's slippery!" She gripped Lucy for balance. As the wheel turned and they reached the top, their platform shifted and they needed to reposition on the other side of the board. They straddled the seat, and as the wheel continued to turn, they flipped to the other side. The castle floor quickly approached and Rosie knew they could not miss their landing.

"Jump on the count of three," Rosie squeezed Lucy's hand. "One, two, THREE!"

Their timing was perfect and with arms and legs flying, they landed on the castle floor in a puddle.

"We did it!" Lucy exclaimed. She jumped up and down.

Rosie couldn't celebrate yet, because she was worried how Nathan was going to manage Maddy on the wheel.

She saw them coming over the top and Nathan held Maddy in his lap. Maddy had her nose in the air trying to avoid the water splashes. At the top of the wheel, Nathan flipped around on the board with Maddie tucked under his arm. They made it, but now he had to make the jump off the wheel. Nathan held tight to Maddy, and launched.

SPLASH.

They came up short and landed in the river below.

"Maddy!" Rosie ran to the water's edge and held out her hand. Nathan still gripped Maddy tight and Rosie was able to pull them in.

"Are you okay?" Rosie asked as she wiped Maddy's face and looked to Nathan.

Maddy spit out a mouthful of water. "That was fun!"

"I'm fine." Nathan managed. "But I don't want to do that again."

"Thank you, Nathan, for taking Maddy," Rosie said as Nathan blushed.

Nugget came running up to Rosie's feet and drank from the puddle she had formed on the floor. They were all on the other side, and Princess Nilly was overhead.

"We're inside the Grymballia Castle. We must hurry before someone hears us or our invisibility shield disappears." She fluttered off around the corner, and they followed.

The castle had vaulted ceilings and ornate walls adorned with the wonders of nature. Artwork, statues, and sculptures consisted of plants, waterfalls, and rainbows. Mother Nature was honored in every direction. The plants vined down the walls and the air was warm and thick with moisture like a greenhouse. Sunlight poured into the windows to feed the plants.

As they hurried toward the main entrance, a large foyer of space opened up ahead. The royalty of the Grymballian castle greeted visitors with fountains, flowers, and a gold staircase that reflected the sunlight pouring inside. Two Plyrim soldiers stood at the front entryway and their smell overpowered the sweet blooms. Giblet looked at his arms with a furrowed brow and then checked everyone else to assure they were still invisible. Time was ticking.

They ducked behind a column. Rosie whispered, "We need to get upstairs to the King and Queen before we're discovered."

Lucy held up her hand. "But there could be ten soldiers around the corner that we haven't seen yet."

"I'm going to scope it out first." Before Rosie could object, Nathan crept toward the main lobby.

Rosie held Maddy close as images of her falling off the waterwheel recurred in her mind. Maddy wrapped her arm around Rosie's waist and seemed to know that the bad smell meant bad news. Rosie rested her hand on Maddy's shoulder.

Nathan walked out into the open lobby and right past two soldiers standing guard. The Plyrim soldiers stood talking and didn't flinch as Nathan tiptoed past. Nathan reached the far side of the room and they lost sight of him as he disappeared around a corner.

Rosie stepped out past the stairway to find him and finally saw the extent of the castle's magnificence. They hid under a golden stairwell but in the back of the room an arch of gold leaves and cascading vines stretched overtop of two thrones. Fluorescent stones, similar to the ones Princess Nilly gave Maddy, speckled the thrones on a base of gold. One throne was made of a sunflower with its petals open wide, and the back of the throne was adorned with pointed golden leaves. White birch tree stumps formed the legs and Rosie knew it was the Queen's throne. The King's throne had a mushroom cap turned upside down to form the seat while it was secured with four pinecone legs. The back of the throne was made of peacock feathers. She sketched in her journal to remember it forever.

Rosie's mouth opened in awe. She stepped further into the lobby, still no sight of Nathan, and she wanted to see more. A sun hung over the thrones made of different yellow flowers woven into a large ball of light. It was breathtaking. Rosie was determined more than ever to return the King and Queen to their throne.

Nathan appeared and hurried their direction. Giblet suddenly pointed and whispered, "It's wearing off!"

The soldiers still talked and did not yet notice the faint outline of Nathan sneaking by them, but Rosie saw a soldier rub his eyes and look over his shoulder. Rosie waved her hands frantically to get Nathan to run, and he sprinted and dove under the stairs where they hid. The soldier looked around and repeatedly rubbed his eyes, but then resumed his conversation.

"That was close." Rosie whispered to Nathan. "Your invisibility was wearing off. He almost saw you."

Nathan panted. "We need to hide. It's not safe here."

Rosie directed everyone down the hallway from which they arrived, and they ducked into an empty room.

"What do we do now?" Lucy asked.

"What did you see in the lobby?" Rosie asked Nathan.

"There are two soldiers in front of the castle and the two in the lobby. The King and Queen's thrones are there, so that's where we need go. The stairwell to the dungeon is around the corner." Nathan rested his hands on his knees to catch his breath.

Rosie stared at the ceiling and tugged on her hair. "We have to get past the guards out front and in the lobby, get upstairs, defeat a monster Fligarian, and then bring the King and Queen to their throne." Rosie tilted her head and looked to the others. "No sweat, right?"

"I have an idea." Nathan's eyes twinkled. "But it's dangerous."

Chapter 11

Nathan's plan was dangerous. He envisioned a special tool to battle the Plyrim soldiers and Fligarian that magical creatures could not detect. He had an idea, but he needed to leave Grymballia and wanted Giblet to come with him so he could get through the portal to the outside world.

"We need something to fight with. The soldiers are too big, so we'll have to outsmart them." Nathan grabbed his backpack and prepared to leave.

"What are you going to get?" Lucy asked.

"I have an idea, but I'm not sure I can pull it off."

Rosie paced back and forth. If they had weapons, they were still outnumbered by Plyrim soldiers. "What if we free the prisoners from the dungeon first, so that they can help us fight to rescue the King and Queen?"

Giblet's face lit up. "That's a great plan. I know many of the captured would fight." His face softened. "Especially my brother, Goblet."

Rosie nodded. "Okay, then while Nathan and Giblet are gone, we will free the prisoners."

Lucy looked to Rosie as if she was a lunatic.

"I don't want Giblet to leave." Maddy ran to his side.

Giblet smacked his tail and after the sparks settled, he held a bouquet of flowers for Maddy. "I'll be back. Can you help Rosie while I'm gone?"

Maddy buried her face in the colorful blooms and nodded to Giblet with a tear in her eyes.

"It's a good plan, Rosie. We'll be back soon." Nathan said.

Giblet smacked his tail again and a castle window appeared. A warm breeze and sunshine poured inside. "You can't ride the waterwheel out, Nathan, but you can go out the window." Giblet pointed, and then squeezed through the crevice through which he and Nugget had crawled inside.

Without hesitation, Nathan removed rope from his backpack and threw one end through the window. He felt Giblet's tug on the other side that it was secured. Nathan crawled up the rope and barely grasped the windowsill before his tired hands gave out. Rosie knew she could not make that climb, especially with Maddy and Nugget. Nathan disappeared through the window and out of the castle.

Princess Nilly called for a Blim Bird to meet Nathan and Giblet in the forest, and then to escort them to the caves while hiding in rainbows along the way.

Rosie looked to her crew: Princess Nilly, Maddy, Nugget, and Lucy. They would need more help to rescue the prisoners. "Princess Nilly, are there others in Grymballia that have not yet been captured that might be willing to fight?"

The princess flitted in circles. "Yes! I know of a few villages we can visit."

They decided to leave the castle and gather more troops for the rescue. They needed another way out, because they could not ride the waterwheel out of the castle without diving underwater for extended periods of time.

Blim Birds cannot fly close to the castle or they would be spotted, and the opening next to the wheel was not big enough for anyone but Nugget to squeeze through. And Rosie knew from gym class experience that she would never make it up the rope.

"I have an idea." Princess Nilly flew close. "Everyone huddle together."

The princess' twig arms waved and her wings buzzed to a blur of pink and red as she closed her eyes. She clapped her hands, and suddenly, they were enclosed inside of a large bubble.

Lucy giggled as she looked around at the clear wet walls. "This is so cool."

Rosie scooped up Nugget so her toenails would not puncture the bubble, and she pulled Maddy close as their bubble lifted off of the castle floor and started to float. It rose high above the ground and Rosie's stomach did a cartwheel as she looked far below. They drifted toward the waterwheel and the small opening leading to the outside.

"Are you steering this thing, Princess?" Lucy looked down at the churning water below.

The princess flew around inside the bubble and guided it through the opening of the castle and into the sunlight outside. They glided in the breeze and over the treetops, but then a gust of wind thrust them forward.

The princess tumbled inside the bubble as it flew out of control. Rosie lost hold of Nugget and tried to grip Maddy as they sailed into a cluster of trees.

Maddy squealed, "Wheeeeee!"

Lucy screamed. "I don't want to die in a bubble."

The wind died down and the bubble slowed, but Nugget jumped off of Rosie's lap and barked at all the excitement.

Her toenails popped the bubble.

They tumbled to the earth with a thud.

"OOWWW!" said Maddy as she rubbed her bottom.

"Are you guys okay?" Rosie asked.

"I twisted my wrist," Lucy said. "But I'm fine. I prefer Blim Birds any day." She rubbed her arm.

"Maddy, are you hurt? Can you walk?" Maddy was not screaming which Rosie felt was a good sign.

"Dumb, Nugget." Maddy yelled. "You popped our bubble."

Rosie scratched Nugget's ears. Poor dog had been flying on Blim Birds and inside of a bubble surrounded by strange creatures. "Nugget's had a big day."

They had landed in the middle of the forest and could no longer see the castle.

"This way, my friends," Princess Nilly urged as she fluttered off between the trees. "The village of the Larmox is nearby."

"Larmox?" Lucy whispered.

Rosie shrugged. "We need as much help as we can get. Geez, the princess is really moving!"

They dodged branches and jumped over small bushes to keep up with Princess Nilly as she flew overhead. Panting and sweating, Rosie was about to suggest a rest when she spotted a clearing ahead. Nestled in the dense trees was a tiny village.

Small houses scattered the clearing and were made of different forest items. One house was a hollowed out coconut shell with a pine needle door waving in the wind. Another house was made of white birch bark plastered together with sap, and its roof was covered with acorn tops to channel away the rain. Another house was made of cattails bound together for plush, soft walls with a billowy roof of milkweed. It was Rosie's favorite.

Princess Nilly sang a song and the Larmox peeked out of their houses.

From behind the pine needles of the coconut house drifted a plump worm-like creature with a top hat. He was purple with green stripes and no arms or legs. He had big brown eyes and friendly smile.

Rosie nudged Lucy and whispered, "How is he moving without legs?"

Lucy pointed.

A tiny pair of green wings flapped high on his back.

"Hello, Princess Nilly. It's an honor to have you to our village." The Larmox bowed to the princess.

"Hello, Sammy. It's good to see you and see that your village is safe. We brought Rosie from the outside world to help us save the King and Queen." The princess motioned to Rosie.

Rosie waved awkwardly as she shifted on her feet. "We're going to free Grymballia from the soldiers so you'll no longer have to live in fear."

Sammy bowed to Rosie. "We're so grateful you have come to help." He looked over his shoulder. "Did you hear that, Jojo? Did you hear that, Henry?"

Henry was a chubby worm and a solid brilliant blue. Despite his bigger size, his wings were the same as Sammy's causing him to hover close to the ground as he flew. His face scowled. "How can we trust them, Princess Nilly?" Henry pointed to Nugget. "That one looks like he's going to eat us all."

Rose knelt down. "Henry, this is my dog, Nugget. She won't hurt you." Henry just glared at Rosie.

"Henry, our friends are to be trusted. We need their help, and they need yours as well." Princess Nilly then looked to Sammy and Henry.

"OUR help? What can we do?" Sammy asked.

"We're gathering helpers to free the King and Queen. If we free the prisoners in the dungeon, we will have more firepower for their rescue." Rosie explained.

A young Larmox girl appeared from the cattail hut. Jojo was a petite Larmox with spunk. She was bright pink with red polka dots and had pig-tails on her head, while the other two Larmox were bald. Her wings were crimson red as were her lips. She flew to the center of the conversation. "I'm in."

"Jojo, you have no idea what you're doing. This could be a very dangerous mission." Sammy scolded.

86

"I'm tired of hiding from soldiers and its time we started to fight for what we believe in. I'm strong, Dad, and I'm going." She looked right at Sammy. Rosie thought that Jojo would have stomped her foot for emphasis – if she had one.

Rosie and Lucy grinned. Girl power.

"Thank you, Jojo," Rosie said. "But we could actually use everyone's help."

Sammy and Henry looked at each other.

"The soldiers have stolen Fred and Bucktooth from our village already, and we live in fear daily. If Jojo is going, then I am too." Sammy tipped his top hat.

"Thank you." Rosie introduced everyone in the group. Maddy kept trying to pet the Larmox and Rosie had to gently pull her away.

"Henry, will you help us?" Lucy knelt down to Henry. He turned his eyes away, but then met her gaze.

His chubby cheeks blushed. "Okay," he grumbled.

"We will fight together." Sammy slapped Henry on his back.

The Larmox could burrow underground like an earthworm, but could also fly overhead like a butterfly. They stirred up the soil to help the flowers grow in Grymballia. If captured, they could release an instant layer of slime over their body. Henry had slipped out of a Plyrim soldiers grasp the week prior to escape.

Their team was growing.

Chapter 12

Rosie's confidence grew as the Larmox joined their group and they headed toward the castle. Henry thought it best if the Larmox traveled underground while the rest of the group went on land. If they split up their large group, then they had less of a chance to be spotted. The Plyrim soldiers could not detect the Larmox underground. Sammy tipped his hat to Rosie, flapped his wings, lifted off the ground, and then plunged into the earth leaving a mound of soil in the grass. Henry and Jojo dove in a similar manner while Rosie's group stared at the dirt piles where they had disappeared.

"Where did they go?" asked Maddy.

"They're digging their way to the castle, and we're going to meet them there." Rosie answered.

Nugget sniffed the dirt piles that the Larmox had left behind, and Rosie chuckled to herself that Nugget was probably very confused about this world.

As they walked toward the castle, they hid behind trees, and ducked into bushes to avoid detection. Rosie hoped Nathan and Giblet were safe and would be back soon, but she couldn't worry now, she had to keep her team safe.

Maddy slowed down and plopped onto a rock to rest. "Rosie, I'm tired of walking." She looked up with droopy eyes.

Rosie was impressed that Maddy had kept up for this long. "We have to keep moving, Maddy. We can't let the Plyrim soldiers find us and we have to meet Nathan and the Larmox at the castle."

Maddy's head fell and Rosie could see her tears beginning. Rosie knelt down. "Do you want to hop on my back?"

"Really?" Maddy crawled up immediately. "You never want me to ride piggy-back."

Maddy hopped on and nuzzled her head into Rosie's back, and her curls tickled Rosie's ear.

Rosie couldn't help but notice how tiny Maddy was, and how cozy she felt on her back. Lucy dragged her feet as they walked.

After walking a short way, Rosie's stomach gurgled. "Let's stop for a snack. We're going to need our energy to free the King and Queen." She found Gatorade and crackers in her backpack.

Maddy jumped down. "I have two Dum-Dums left Rosie!" Maddy smiled and pulled them out of her pocket, but then pulled them to her chest. "But I want to share them with Giblet. They're his favorite."

Princess Nilly drifted down from above with her eyes shining. "Girls, I have a treat for you. We don't have candy in Grymballia, but that doesn't mean we don't have sweets."

They followed the princess flew through the trees, and soon the forest opened up into a grove of colorful fruit trees. Trees stretched in all directions covered with pears, apples, cherries, peaches, and oranges. Rosie saw gardens and fields continue into the distance packed with growing vegetables.

"Is this magic fruit?" Lucy asked as she plucked an apple the size of a cantaloupe.

"Help yourselves." The princess smiled. "The Larmox stir the soil to fertilize our crops and it makes our fruits and vegetables bountiful. One piece of fruit can feed a whole family." The princess rested on a tree branch. "It's also why Plyrim wants Grymballia. They abused their own land and crops no longer grow."

"I don't like fruit." Maddy crossed her arms over her chest.

"Maddy, don't be rude!" Rosie was embarrassed. "You've never eaten Grymballian fruit. Just try it." Rosie appreciated her mother and how she battled with Maddy at every meal.

Rosie pulled a humongous pear off the nearest tree and it was so heavy that she almost dropped it. It had smooth, yellow skin and it smelled so sweet that her mouth watered. She handed it to Maddy.

Maddy reluctantly grabbed the pear with a frown on her face, but after one sniff of the juicy fruit, she sank her teeth into the soft flesh. Juices exploded down her chin and her eyes lit up with fireworks. "YUMMY!"

Maddy sat in the grass and devoured the pear, while Rosie explored to find a raspberry bush loaded with ripe berries. One berry took three bites to eat, and Rosie moaned as the sweetness filled her up. Delicious. Lucy plucked off a peach and dove in.

"Are you going to eat, Princess Nilly?" Rosie asked.

The princess waved her hand while she hovered overhead to watch for soldiers. Even Nugget nibbled on a blueberry. They sat on a blanket of grass as sticky juices covered their faces.

Rosie noticed a flicker of color move in the bushes.

"What was that?" Lucy's eyes grew big and she stood up quickly.

Rosie crept toward the bushes and saw five brightly colored butterflies hovering over a cluster of flowers.

Lucy peeked over Rosie's shoulder. "What are they? They're not butterflies."

Rosie leaned in and realized that the large colorful wings contained many small eyes – that were watching them. The body was furry like a miniature squirrel and they had a bushy tail. She did not see a mouth.

"Hello, Rosie." The butterfly-creatures flapped their wings rapidly to form sound.

"H- h- hello," Rosie responded in shock that all the creatures knew her name. "What are you?"

"We are Glaperia." The largest of the Glaperia fluttered with fluorescent blue wings. "We can see for miles and also see through objects."

Rosie looked to Lucy.

Lucy stared. "That is so cool to have x-ray vision! But how are you talking without a mouth?"

"Our wings beat at a frequency to allow our thoughts to speak." The Glaperia hovered close to Lucy.

"What's your name?" Maddy had walked close and still sucked on her juicy pear.

"I'm Xena, and my sisters are Nina, Dina, Trina, and Patsy," she said.

"Patsy?" Maddy made a funny face. "That doesn't rhyme with the rest?"

Rosie elbowed Maddy.

"Patsy is . . . different. Her wings beat to a different drum." Xena giggled.

Rosie easily figured out which Glaperia was Patsy. Unlike her sisters, she sat instead of fluttered, and all twenty-four of her eyes scowled at Xena. She was taller and thinner than her sisters and her fur was midnight black. Patsy's wings popped with multiple bright colors, and they started to flutter as she glided toward Rosie.

Patsy focused on Rosie and her wings beat as rapidly as her speech. "Where are you guys going? Can I come with you?" She kept her back to Xena. "I'm so bored."

Rosie glanced between Patsy and Xena. "We're going to the castle to save the King and Queen."

ALL of Patsy's eyes lit up.

Rosie tilted her head. "We honestly could use your help." Rosie was afraid to get Patsy in trouble from Xena. "If Glaperia can see through objects and for a long distance, that could be useful. You're welcome to join us." Rosie paused. "Actually, all of you could join us." She looked to the Glaperia sisters.

"I'm going," Patsy said without hesitating.

"It sounds dangerous. What's your plan?" Xena asked.

"The Larmox are helping, and our friend, Nathan, is bringing tools to help us fight. If we free the prisoners in the dungeon, then we'll have more help to rescue the King and Queen." Rosie felt stronger about their plan the more she said it.

"You'll need as much help as you can get. My sisters and I will join you also," Xena said.

Rosie bounced up and down. "That's awesome. Thank you." Rosie glanced to the princess in the treetops. "We should probably keep moving toward the castle."

Xena motioned to her sisters. "Let's fly."

Patsy sped ahead of her sisters as they flew from shrub to shrub. Maddy skipped along the trail now energized by her pear. The golden dome of the castle appeared ahead, and they headed toward the meeting point.

And then they smelled trouble.

The princess zipped down from above yelling, "Soldiers!"

Nugget barked, Maddy screamed, and Rosie grabbed them both and dove under the bushes. The Glaperia flew high out of sight and Lucy ducked next to Rosie. The Plyrim soldiers cloud of black smoke approached, and Rosie quieted Nugget as they lay on their stomachs.

The disgusting soldiers patrolled close and the stench was overwhelming. Rosie gagged. Nugget squirmed in Rosie's arms and Maddy's face filled with fear as she watched the soldiers.

Maddy whimpered – and the soldiers froze.

92

Rosie panicked and covered Maddy's mouth with her hand, but as soon as she moved, Nugget bolted from under the bush. Nugget ran up to the soldiers and bit their ankles as she growled.

Rosie stared and wanted to scream for Nugget, but it would risk the soldiers finding them under the bush.

"What is this thing?" The Plyrim soldier shook his leg. "It's biting me."

The other soldier snorted. "Must be another Grymballian. Put it in the bag, and we'll throw it in the dungeon with the rest of them."

Rosie's heart raced and her muscles twitched, as she wanted to run out and grab her dog. Nugget in the dungeon! What would happen to her there? Rosie's mind flew through her options. If she ran out, she and Maddy would also be captured and she had nothing to fight off the Plyrim soldiers. She had to stay quiet.

She watched the soldiers take Nugget by the scruff of the neck and stuff her wiggling and growling into a bag. Tears flowed down Rosie's cheeks as she watched the soldier throw the bag over his shoulder and walk off in a cloud of smoke.

Nugget was gone.

Rosie realized she was still holding Maddy's mouth when the coast cleared. They crawled out from under the bushes.

"Rosie, they took Nugget! We have to go get her." Maddy pulled on Rosie's shirt and pointed toward the disappearing cloud of stink.

Rosie knew she had to be calm. She had to act like a grown up, but she had never felt more like a child. She was scared and wanted her dog. Lucy looked at her with wide, worried eyes and then gave Rosie a hug.

Lucy raised her fist. "We'll get her back and show those smelly punks they shouldn't mess with us," Lucy yelled.

Rosie stood tall. "We'll save her with the other prisoners." She faltered. "Nugget's my best friend." She looked into Lucy's eyes and smiled. "My other best friend."

Princess Nilly patted Rosie's shoulder. "Rosie, the other Grymballians will take good care of Nugget in the dungeon."

Rosie looked to Lucy and the princess, Maddy and the Glaperia. "Let's get to the castle and meet up with the Larmox. Hopefully Nathan and Giblet will return soon, and then all of us will save Nugget and the others." Rosie's anger made her stronger and determined. "And then we will free the King and Queen to save Grymballia. Let's do this."

Chapter 13

Rosie's team waited for the Larmox to arrive by a large walnut tree. She paced back and forth checking over her shoulder for Nathan while watching for soldiers. The Glaperia stayed high in the treetops cautiously, except for Patsy who hovered close to Maddy and Lucy looking for adventure.

Xena called down to Rosie, "We can see through the trees and will know if a soldier approaches. All is clear."

Rosie gave Xena a thumbs-up and tried to stay quiet to avoid drawing attention from soldiers.

Suddenly, the ground opened up at their feet and three heads popped out of the dirt. It was the Larmox. They surfaced and Sammy put his top hat back on.

Jojo flew in circles around the team. "Did we miss anything? What's happening? Any soldiers?" She chatted fast with excitement.

"We've had a few changes – some good and some awful. The good news is that the Glaperia are helping us." Rosie introduced them all, and Patsy and Jojo immediately stayed by each other's side.

Sammy said, "Wonderful. The more help we get, the better the chances Grymballia will be free again."

"What's the bad news?" Jojo's pigtails bounced as she fluttered.

"The Plyrim soldiers took Nugget to the dungeon with the other prisoners." Rosie hoped they would care that Nugget had been taken.

"Oh, no!" Jojo exclaimed and her chubby worm body shook. "Then what are we standing around for? We need to go after them."

"I agree," said Patsy as she fluttered her wings faster to talk. "We'll get Nugget back."

Rosie wanted to cry as her heart swelled with the warmth of her new friends. She watched Jojo and Patsy immediately bond as two young girls trying to rebel against their parents. They jabbered continuously.

"Princess Nilly, can you see Nathan and Giblet yet?" Rosie whispered up to the treetops.

"Not yet," Princess Nilly said. "But the Blim Birds know they can't fly this close to the castle or they will be spotted, so Nathan and Giblet would come on foot the rest of the way."

Rosie paced. Nathan and Giblet had been gone a long time and the sun started to drop into the afternoon sky. Would they have enough time to rescue the dungeon prisoners, save the King and Queen, free Grymballia, and then get home for supper?

Xena fluttered to a shout. "Something's coming! Everyone hide."

Rosie grabbed Maddy's arm and they dove under the bushes. Even though Patsy could fly high above to hide, she hovered low with Jojo. Rosie didn't smell soldiers?

From under the bush, Rosie glimpsed Nathan's hiking boots and Giblet's potbelly. They were back!

They rushed out from under the bushes and Nathan leapt backward grabbing his chest. Their initial fear turned to relief when they recognized their friends. Nathan's nice shirt was covered with dirt and sweat, and he looked exhausted with a huge backpack full of gear.

"Nathan, I'm so glad it's you. Are you guys okay?" Rosie ran to his side.

"We're fine, but that's a long way home and back. How are things here?" He looked around at the large group as he pushed up his glasses.

"The soldiers took Nugget." Rosie filled Giblet and Nathan in on everything.

Nathan turned to Rosie with a sad look in his eyes. "I'm so sorry, Rosie. But we'll get her back. I promise."

Rosie suddenly had no doubts in her mind that she would see Nugget soon.

Rosie introduced Nathan and Giblet to the Glaperia and the Larmox. They had gathered a good team, and they now needed a plan. "What did you get at home, Nathan?"

"I've been working on a plan and I think we can do this." Everyone huddled in to listen.

"Did you find something to fight the soldiers?" Rosie questioned. "They're so big and there are so many. And then we have to fight the fire-breathing Fligarian in order to rescue the King and Queen."

Henry shook his head. "Impossible."

Nathan grabbed his backpack. "I have weapons – sort of. They should not be detected by magic because they won't recognize them as a threat." Nathan looked skeptical. "It's all I could come up with on short notice."

Nathan pulled two large thermoses coated with frost out of his backpack. He then pulled out two six-pack bottles of Diet Coke.

Lucy laughed. "Are you thirsty?"

Nathan ignored her and pulled out a dozen packages of Mentos candy.

Rosie knew immediately. "Nathan, you're a GENIUS!"

Lucy raised her eyebrows at Rosie and confusion filled her face.

Rosie said, "It's our science experiment. The Diet Coke and Mentos will make an explosion when mixed together."

Lucy nodded as she remembered the sticky classroom, and Rosie ran up to Nathan and hugged him.

His face flushed. "Thanks, Rosie."

"What's in the frosty thermos?" Rosie grabbed the cold container and it froze her fingertips.

"Liquid nitrogen."

Perfect. Rosie looked upon her team of confused faces as they saw Coke, candy, and an icy thermos. She explained that they could make explosions and freeze their opponents. A ripple of excitement ran through the team as Sammy and Henry pointed at the thermos, and the Glaperia flew closer to examine the Mentos.

Rosie and Nathan devised the plan. First, they would rescue the Grymballians from the dungeon to gather more help, and then they would battle the Fligarians to free the King and Queen. Rosie's mind drifted to Nugget and she hoped he was safe.

They headed for the waterwheel to sneak inside. Giblet slipped through the small passage beside the wheel, while the Glaperia, Larmox, and Princess Nilly floated overtop. If Nathan got the Mentos wet, they would dissolve and not work with the Diet Coke, so he rode alone on the wheel while constantly lifting the backpack to avoid the water. He landed safely on the other side – a little wet, but the backpack was dry.

Maddy rode with Rosie and they made the final leap easily. Lucy arrived last. The stench of Plyrim soldiers filled the air, and they had to reach the dungeon door on the other side of the castle quietly and carefully.

Rosie whispered, "Most of the soldiers guard the front door so let's try finding a back route." She pointed down a hallway.

Princess Nilly flew high with the Glaperia, and the Larmox burrowed through the plant-boxes along the corridors. Rosie, Nathan, Maddy, and Lucy hugged the walls and tucked behind potted flowerpots when possible.

"Something's around the corner up ahead. I can see it moving through the wall." Trina the Glaperian swooped down to warn them.

"Thank you, Trina!" Rosie was thankful for Trina's special x-ray eyes. They ducked behind a fountain between two rose bushes. They smelled soldiers.

Two soldiers sauntered by talking in snorts and grunts. "The dungeon's almost full."

The other soldier responded. "We'll just get rid of some prisoners to make room for more."

"I don't know why we're keeping them anyway. Let's start with that new prisoner that makes all the noise." The soldier chuckled an evil laugh. "I can't wait to quiet that bark of his."

NUGGET. Rosie thought as she squeezed Lucy's hand.

The soldiers disappeared around the corner in their smelly cloud. Nobody had been spotted. Nathan glanced at Rosie with sympathy in his eyes. Giblet looked pale and shaky because his brother, Goblet, was in the dungeon, too.

Xena flew down, "It's all clear."

Rosie marched ahead of the pack and saw the dungeon door. She whispered, "I'm coming, Nugget."

Chapter 14

Tension filled the air as they approached the dungeon. Maddy hugged Rosie's leg and Lucy tugged on her shirt to stay close. The dungeon door was made of solid wood with ornate tree bark and branches woven into a sturdy frame. Silver intertwined with jewels scattered the exterior giving the dungeon a beautiful appearance. Rosie knew that beyond the beauty lie soldiers, prisoners, and a battle.

"Patsy, can you see through the door and check if anything waits on the other side?" Rosie continuously glanced over her shoulder for oncoming soldiers.

Patsy was thrilled to help and hovered in front of the door. She shifted her body back and forth as her winged eyes searched every inch of the door.

"There's nothing on the other side. Let's hurry." Patsy looked to her sisters for approval and Xena nodded with a smile.

Lucy and Nathan heaved the door open and the overwhelming stench of the Plyrim soldiers almost knocked them off their feet. A spiral staircase of flat stones wound downward into darkness, and Rosie paused to let her eyes adjust. The Larmox forged ahead – they were used to burrowing in darkness. The Larmox hopped their plump worm-bodies down the steps instead of flying in order to conserve their energy.

Rosie whispered to the Larmox. "Sammy, let us know what you see."

The group inched forward and Princess Nilly floated above in the lead. Xena and her sisters assured that they saw nobody waiting ahead.

A bark rang through the dungeon.

Nugget!

Rosie recognized Nugget's bark and knew she was mad. She hoped they reached her before the soldiers did. Rosie clenched her teeth as she replayed the Plyrim soldier's conversation in her head to *get rid of him*.

"NUG-" Maddy yelled, and Rosie quickly clamped her mouth shut.

They froze. Did the soldiers hear Maddy scream? Rosie removed her hand from Maddy and shushed her with a finger to her mouth. Maddy nodded.

Rosie whispered, "I hear her, too. We'll save her."

At the bottom of the stairs, Princess Nilly and the Larmox peered around the corner to examine the dungeon.

The princess explained in a whisper, "A large chamber is ahead and the prison cells are on the other side of the room. The Ackly vines are coated with thorns and form the prison cells. Each thorn contains poison to put you into a deep sleep. The soldiers wear armor to protect them from the poison."

Rosie stared dumbfounded at Princess Nilly. How would they get past the Ackly vines even if they got past the soldiers? "Do the Ackly vines have any weaknesses, Princess Nilly?" Rosie asked.

"The Ackly vines are sad and lonely from being trapped in the dungeon for years. The sound of music changes their behavior."

Nathan began to load their weapons. He handed Rosie and Lucy two bottles of Diet Coke and one box of Mentos each. The Larmox stood ready to fight, and the Glaperia huddled close. Rosie kept Maddy by her side so she could help free the prisoners.

"Maddy, don't touch the Ackly thorns or they'll make you sick. We have to get past the stinky soldiers, okay?"

"I'm ready, Rosie." Maddy looked into Rosie's eyes and nodded.

Rosie thought that Maddy suddenly looked grown up.

The team was in position. Patsy peered through the wall to see what was ahead. She flapped her wings and concentrated her many eyes.

"Three soldiers sit at a table playing cards. There are five prison cells sealed with spiked vines and every cell is packed with Grymballians." Patsy turned to Rosie with a hint of panic.

Rosie pulled everyone close. "Nathan, Lucy, and I will go for the soldiers. Giblet, can you distract them? Then we can surprise attack while the Larmox and Glaperia help Maddy free the prisoners."

Giblet stepped forward. "I have an idea." He paused and looked to Princess Nilly. "Princess, you must stay out of danger because Grymballia needs you."

Maddy threw her arms around Giblet. "Be careful. I love you, Giblet."

Giblet's green color glowed and he turned to Maddy. "I love you too, Maddy." Maddy's love gave him confidence as he bolted forward.

"Everyone be ready." Rosie looked at her team and was met with smiles and nods of solidarity. They crept forward into the dungeon chambers and saw Giblet lurking in the shadows along the back wall. The smell of Plyrim soldiers burned Rosie's eyes.

Trina whispered from high above. "The soldiers don't see Giblet."

Giblet gathered small stones and a pile of dirt. He crept into the middle of the chamber and mounded the dirt and stones on the floor. The soldiers grumbled and grunted over their cards and did not see Giblet as he raised his arms over his head and smacked his tail on the floor three times. Sparks flew. A pile of food appeared where Giblet had mounded the dirt and stones. The smell of roast beef and yams caused the soldiers to turn toward the pile and scurry toward the succulent food.

One soldier yelled, "Where did this come from?"

The soldiers looked around wildly but could not see Rosie and the team hiding in the stairway. They also did not see Giblet tucked against the wall. Drool ran down their faces as they hovered over the food.

"I don't care where it came from," said one soldier as he grabbed a goblet of water and guzzled. "We haven't had clean water in Plyrim for months. I'm drinking it." The soldier downed the water in three gulps. He grabbed a handful of beef, burped, and rubbed his belly.

The other two soldiers realized they were missing out and then dove into the food. They shoved turkey legs, bread, and mashed potatoes into their mouths all at once and then got on their hands and knees to slurp more food.

Rosie knew it was time to act. They opened the lids of their Diet Coke and held one in each hand. They charged!

With their heads buried in food, the soldiers did not see them attacking from the stairwell. Rosie's legs wobbled and heart flipped as she scanned the dungeon. Nugget barked frantically when he saw Rosie and caused a soldier to look up.

"Attack! We're being attacked!" The soldier tried to get up but slipped on a pile of grapes.

Rosie acted fast. She pointed her Diet Coke bottle at the soldier and dropped in two Mentos. Whoosh! The eruption immediately burst into the soldier's face and he fell backward onto the other soldiers.

Maddy, the Larmox, and the Glaperia scurried to the prison cells. Maddy ran to Nugget's cage.

"It's okay, Nugget," she whispered. "We're here to rescue you."

Ackly vines and thorns jutted in every direction.

"How are we going to get them out?" Nina exclaimed.

Jojo moved toward the vines and slimed herself until she was coated. When she was slippery enough, she slipped through the vines and into the cell. She avoided the poisonous thorns and gathered the Grymballian prisoners.

"Gather round, we're here to rescue you but we need your help." Many different creatures nodded their head with excitement and surrounded Jojo. Nugget sniffed Jojo and licked her head.

The fallen soldier was angry. He jumped back to his feet and Lucy hit him with another Mentos bomb. He slipped on the sticky soda and fell into the pile of food. An apple stuck to the large yellow tooth jutting out of his face. The other soldiers tried to get up but also slid on soda and tripped on each other.

The Glaperia had an idea. "Music! Princess Nilly said the Ackly vines love music. If we sing, maybe they will release their wall of thorns."

Maddy took charge. Her voice rang loud and clear as she began to sing.

Bright sunshine and rainbows,
Playing children to and fro,
Flowers, trees, and green grass,
Springtime's here at last.

Her sweet voice echoed off the dark, dungeon walls and everyone paused in the chaos. Rosie recognized the song from Maddy's school concert. The Ackly vines missed the sunshine and would love the song. Maddy sang again:

Bright sunshine and rainbows,
Playing children to and fro,

Flowers, trees, and green grass,
Springtime's here at last.

Everyone joined in and a chorus of voices filled the room. Princess Nilly's beautiful voice rang loud as Christmas bells, and the Glaperia's wings flitted wildly to produce sweet tones from their body. The Larmox sang loud and clear, and then – every Grymballian prisoner sang.

The soldiers groaned at the heavenly music, but the Ackly vines – shifted. Thorns fell to the floor and the leaves moved. Everyone sang louder.

Bright sunshine and rainbows,
Playing children to and fro,
Flowers, trees, and green grass,
Springtime's here at last.

The vines unwound from each other and the prisoners were free. Nugget ran toward Rosie, but a soldier jumped into Rosie's face. Nugget bit the soldier's leg and stopped the soldier's attack. Rosie had time to reload Mentos into a Diet Coke and fire. The soldier rubbed his eyes and fell backward. Nathan battled another soldier but was out of Coke, while Lucy had knocked her soldier to the floor. All three soldiers lie on the floor stunned and staring at their attackers.

The Grymballian prisoners filed out of their prison cells and surrounded the soldiers. The smelly beasts suddenly had fear in their eyes as they realized they were defeated.

A Grymballian creature approached the soldiers. He looked like a smooth, bright yellow ball with tiny arms and legs and with a sharp nose jutting out of his little head. He climbed on top of a frightened soldier's chest. The creature

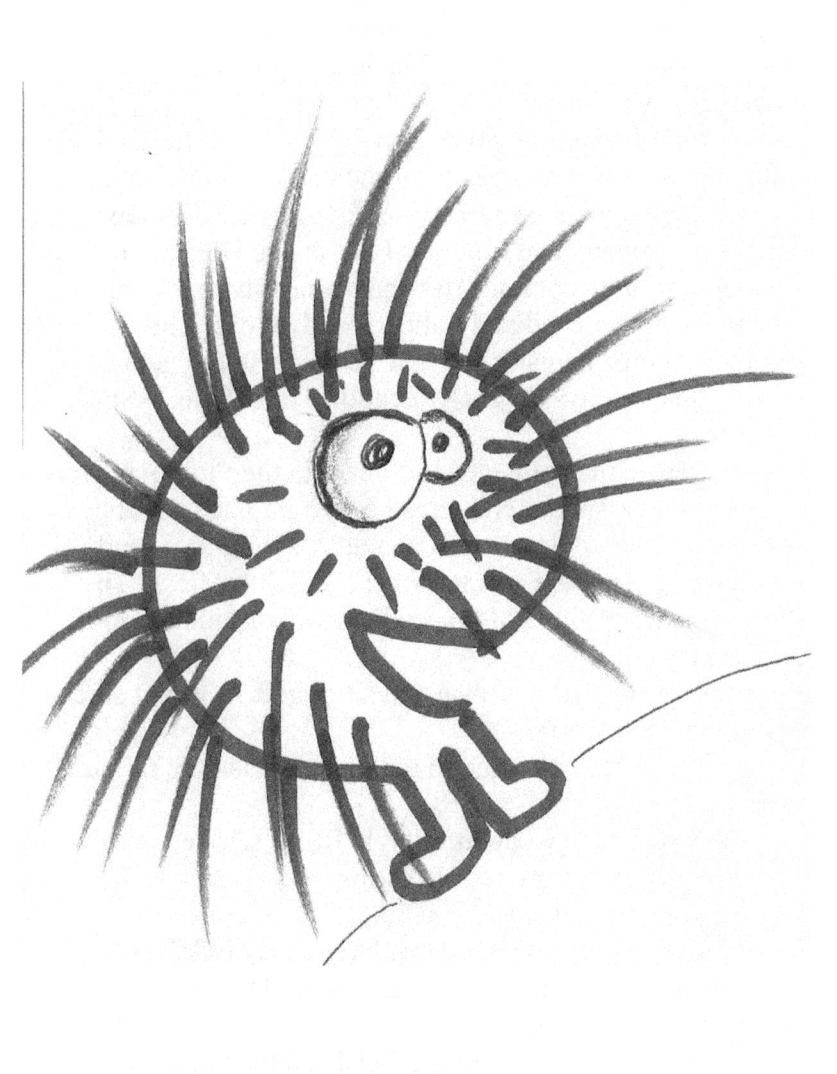

jumped into the air, and when he landed, his body was covered with spikes.

He raised his hand in the air. "On behalf of my fellow Plumpians, you must pay for what you have done. We have suffered without food, water, and our families for weeks." Two more Plumpians rolled to his side. "Rufus and Dennis, you take those two." The leader pointed to the other fallen soldiers.

"Got it, Spike!" said either Dennis or Rufus as they each hopped onto the chest of a soldier.

Rosie looked around to see if anyone else knew what was happening. With only Lucy's one Diet Coke weapon left, they couldn't risk the soldiers warning the others. Rosie watched and pulled Maddy close. The soldiers grunted and spit at Spike as they tried to get up.

"Okay, boys, begin 'Operation Toxic Stab.'" Spike hollered.

Before the soldiers could react to the Plumpian war cry, the yellow spiked creatures buzzed into action. They rolled over the soldier's chest and across their arms and legs leaving puncture wounds over their slimy flesh. The Plumpians hopped off of the soldiers with a satisfied look on their faces.

Lucy looked to Rosie and shrugged. "I'm not sure poking will stop them."

The soldiers started to get up again as they rubbed their wounds.

"Load your weapon, Lucy!" Rosie stepped back.

Spike didn't move. He calmly waved and said, "Wait. Give it a second."

The soldiers started to wobble and couldn't find their feet as they tripped over each other. They swayed and could not walk straight.

"Get them to the prison cells!" Nathan ordered.

With great effort of the Grymballians and the rescue team, they herded one Plyrim soldier to the cells. The disoriented soldier could be pushed in the right direction. The Glaperia, the Larmox, and other winged creatures tugged on another soldier's ears to guide him into a cell. Once the two soldiers were inside, they fell to the floor with a thud. The Plumpian poison had taken its full effect and they were asleep.

One Plyrim soldier remained. He fought off Rosie and Lucy as they tugged on his arms. His poison was not as effective and he stood strong. Just then, Giblet was at their side with an identical creature. His brother, Giblet! They beat their tails on the floor three times in unison and sparks flew. The soldier's feet lifted off the ground as he floated into the air with no control.

"Push him toward the cell!" Rosie told Lucy.

The soldier turned and spun in the air and they pushed him toward the prison cell. His face turned green with sickness as he spun in circles and then crashed to the floor.

"Now what?" Lucy asked. "How are we going to keep them inside the cells?"

Princess Nilly beckoned for silence with the wave of her hand. Everyone quieted as she sang. Her song relaxed Rosie and calmed the crazy room filled with yelling and noise. The Ackly vines rose off the floor in response.

The princess spoke, "Vines of Ackly, we need your help. You're Grymballian family and you've been trapped in the dungeon for years. I'm the current ruler of Grymballia, and I vow that I will grant you freedom from the dungeon if you help us hold the Plyrim prisoners. I will release you to the sunshine and forest if you will help us."

Everyone held their breath as they watched anxiously. The soldiers were still lying on the cell floors, but they were moaning and restless. The vines rose together

107

and intertwined as one. They formed walls and built the barrier needed to hold the soldiers captive. The Ackly thorns stood firm to imprison the soldiers.

"Thank you, my fellow Grymballians," Princess Nilly said. "You're very noble and you'll be rewarded."

Everyone cheered. Rosie looked around to the mass of Grymballians that were rescued. Now to save the King and Queen.

Chapter 15

The mob of Grymballians cheered and Rosie waved her arms to gather them close together. More soldiers awaited upstairs and they still had to battle a Fligarian fire-breathing beast that guarded the King and Queen. Rosie surveyed their assets. They had another six-pack of Diet Coke with Mentos and also the unused liquid nitrogen. The Glaperia – Xena, Dina, Trina, and Patsy fought by their side in addition to the Larmox – Jojo, Henry, and Sammy. Goblet joined his brother, Giblet, in combat, and Rosie had Maddy, Nugget, Lucy, and Nathan. But they needed more help.

"Hello, Grymballians. My name is Rosie." A rumble flowed through the crowd and they pointed to Rosie as they whispered. "My friends and I are here to help save your world, but we need your help. I know you've been held captive a long time and are anxious to return to your families, but are there any volunteers to join our team?"

Rosie scanned the crowd. Immediately, different creatures stepped forward. She glanced to Nathan and the excitement showed on his face.

Rosie continued. "Thank you! Since we have so many volunteers, please step forward and tell us about yourself."

The first Grymballians to step up were Spike and his brothers, Rufus and Dennis. Their Plumpian powers were obvious as the poisoned soldiers snored in the prison cells behind them.

"I'm Spike, and this is Rufus and Dennis. We want to help to free the King and Queen. Operation Toxic Stab pierces our opponent and then puts them to sleep. The bigger they are, the harder they fall." Spike and his brothers stood by Giblet and Goblet and shook their hands.

109

A creature eased forward from the back of the crowd. Nugget ran up to the creature and they went nose-to-nose to sniff each other. Pink tight curls covered her body and she had a small pom-pom tail that wagged when Nugget was near. Then she stood up on her human-like hind feet with six fingers covered with pink hair. She reached out and pet Nugget's head.

Lucy mumbled to Rosie, "Freaky."

"My name is Sid." She had long floppy ears and appeared to be a snuggly puppy – until she smiled. Her mouth was filled with the largest, most ferocious teeth jutting in every direction.

Sid continued, "I'm good at chewing, gnawing, chomping, and crunching. Let me show you." Sid grabbed an empty Diet Coke bottle, and in a flash of fury, plastic bits flew in every direction as the bottle disappeared in a shredded pile.

"Whoa." Lucy clapped her hands. "Those are wicked teeth! Welcome to the team, Sid."

Sid joined the group and Nugget followed her every step. She had found a new friend in Grymballia.

Two snakes slithered up to Rosie. One was brilliant orange while the other was neon green.

"Can we join? We've wanted to be part of an adventure our whole lives." The neon snake spoke.

"Of course," Nathan said. "What are your names?"

"My name is Lila," said the orange serpent. "And this is Kimmie." She pointed to her neon friend. "Kimmie glows in the dark, but I can't."

"Glow in the dark? That's cool." Rosie said as she jotted sketches of each creature and their abilities down in her journal.

"We also have other talents. We hop to great heights, and carry a special poison in our fangs." Lola flashed her two pointed teeth. "We can turn our enemy into water with one bite."

Rosie's mouth dropped open and she stepped back from the snakes. "Wow." The snakes slithered to join the team.

A purple mouse-like creature waddled forward on four flippered feet. Its body was covered in short purple fur with a pair of yellow wings sticking out the side. He dragged a long tail, and one horn jutted out the end of his nose. His deep purple eyes shifted back and forth nervously.

He stammered. "M-m-my name is Tiki. I'm a water Turk. The Plyrim soldiers snatched me from the air as I built my home. I want to help."

"Great, Tiki." Nathan said. "Do you have any talents to share?"

"I-I-I'm the fastest swimmer in Grymballia."

A murmur of respect through the crowd as many nodded in agreement.

Tiki put his head down. "But I don't know how it will help us in the castle."

Rosie tried to cheer up Tiki. "That's an amazing talent and hope to see you swim someday. You should be proud to fight for Grymballia."

He perked up with a smile and then waddled over to join the team. He stood by Jojo and Patsy who chatted quietly. They welcomed him with a pat on the back.

The last creature to volunteer was a brilliant red spider. Rosie's hair stood up on her neck because she did not like spiders.

But this was no ordinary spider.

She had ten legs capped with red sequined boots with silver ribbons. Her torso was covered in red hair that bounced when she walked. Her brilliant green eyes had long black eyelashes.

"Hello, darlings, I'm Priscilla. I will help win this battle," she said boldly.

Lucy nudged Rosie with a giggle.

Rosie said, "Hello, Priscilla."

"I can sling my strong web fifty feet and bring down a soldier." She demonstrated. Priscilla shot a web toward the dungeon wall where it stuck, and then she then slowly walked backward. Her shiny boots strained under the pressure as she pulled, but then a crash broke the silence as the dungeon wall crumpled. A pile of bricks and rock lay on the floor where she had ripped it free. She released her web.

Lucy no longer laughed, but approached Priscilla. "I'll follow you into battle any day!"

No more volunteers stepped forward and many in the remaining crowd would not meet Rosie's eyes. "It's okay if you're not volunteering. We'll do our best to free Grymballia. It's not safe to leave the dungeon yet, so you will have to wait until the battle is over before you can get home safely. Wish us luck."

The remaining Grymballians cheered. Rosie surveyed the team: Nathan and Lucy chatted with the new team members. Maddy hugged Rosie's leg. Nugget sat next to Sid, her pink doggy friend with gigantic teeth. Goblet and Giblet chatted with the round Plumpians: Spike, Rufus, and Dennis. The Larmox: Henry and Sammy admired Priscilla's shiny boots, while Kimmie and Lila coiled up and chatted to each other with their long snake tongues flipping. Tiki sat on his back flippers while he talked to Princess Nilly. The Glaperia: Xena, Dina, Trina, and Nina flew overhead talking about the upcoming battle. Patsy and Jojo hovered together and had been inseparable.

They had a team, so now for the rescue.

Chapter 16

Rosie stocked up on Mentos and grabbed two more Diet Cokes as she prepared to leave the dungeon. Lucy and Nathan also stocked up on supplies while Maddy and Nugget followed closely at Rosie's heels. If they could get past the Plyrim soldiers standing guard in the main lobby, then they could climb the stairs to the top floor where the King and Queen were held hostage. Rosie took two deep breaths and glanced at her team. "Let's do this!"

She climbed the dungeon stairs while others flew, hopped, or leaped up the steps. When Rosie reached the dungeon door, she listened for movement on the other side. Silence. "Trina, can you see anything?"

Trina swooped down and scanned the door with her many eyes. "It looks safe, Rosie." Trina said.

"Thanks, Trina," Rosie said. "You're awesome."

Nathan shoved open the door and peeked through the crack. He saw an empty hallway, so the mob of Grymballians, Rosie, Lucy, Maddy, and Nugget poured out the dungeon door.

They split up into groups as they walked down the hallway so that it would be less obvious if they met soldiers. Nathan, Giblet, and Goblet went to one side, while Lucy, Rosie, and Maddy stayed on the other. The Larmox hid in the flower boxes, Kimmie and Lila hugged the corners of the walls, Nugget followed Sid, and Tiki waddled slowly in the rear on his webbed feet. Priscilla and the Larmox formed a third team in the rear while the Glaperia and Princess Nilly flew overhead.

The smell of soldiers grew to suffocating proportions as they approached the lobby. The sunlight poured into the castle windows but the scene was dark and dirty. Kimmie slithered forward to peek around the corner.

"There are three soldiers patrolling the lobby, but nobody guards the stairs. One team could make a run for the King and Queen."

Rosie shook her head. "We need to stay together."

They whispered a brief plan and then everyone nodded, or waved a wing, or flapped a flipper, or tapped a shiny pink boot to indicate they were ready.

CHARGE!

The soldiers were caught by surprise as they strolled in the lobby. Rosie's heart pounded as she hurried toward the closest soldier with her weapon. His beady black eyes and smashed in face caused her to grimace and almost turn around to run, but instead she popped the lid to her Diet Coke and loaded two Mentos.

Aiming at his flaring nostrils, she hit her target dead center. He stumbled backward as foam flowed out of his nose. Sid and Nugget attacked. Sid's razor sharp teeth grabbed the soldier's leg while Nugget tugged on his ear. The soldier could not stand up to fight them off as black ooze poured from his wounds.

"Way to go, Nugget!" Maddy cheered as she walked a wide circle around to avoid the growling soldier clawing at everyone near.

Lucy flattened her soldier with a Mentos bomb with a direct shot to the forehead. The soldier slipped and fell onto his back and then Tiki jumped on and slapped his face repeatedly with his flippers. It caused no harm, but it was a great distraction for Spike, Rufus, and Dennis to leap on board to use Operation Toxic Stab. The Plumpians pierced his black flesh with tiny holes, and almost instantly, the soldier slowed and his beady eyes became sleepy.

Nathan's Diet Coke bomb hit the third soldier in his eyes, temporarily blinding him, but he did not fall down. Patsy and Jojo and the Larmox swooped down and buzzed around his face. The soldier swung his arms as he tried to

114

shoo away the flying pests, but it caused him to lose his balance and fall with a thud. Lila slithered onto the scene. Rosie was shocked at how Lila hopped into the air and landed on the soldier's head wrapping him in a chokehold. Priscilla smoothed her long hair and then gracefully sauntered over to help. Priscilla shot her web and wrapped the soldier so tight that he was a soldier-mummy, bound and motionless on the floor.

All three soldiers were defeated and temporarily down on the floor. Rosie clapped and pumped her fist to see the soldier-mummy, the sleeping soldier, and last soldier rolling around with black oozing wounds. "Let's get to the stairs!"

Nathan yelled, "Keep your eyes open for more soldiers. Xena, you and your sisters lead the way to watch for danger."

The Glaperia led the way as they rushed the large staircase. They were one step closer to freeing all of Grymballia. The positive energy practically carried them up the steps.

Suddenly, the wounded soldier roared, "INTRUDERS! WARN THE FLIGARIAN!"

Lucy panicked. "What do we do, someone's going to hear him!"

Kimmie and Lila coiled tight and then simultaneously leaped into the air with an amazing height to fly off the staircase and land on the screaming soldier. The Plyrim soldier shielded himself from the wild snakes. Kimmie slithered over his body and then Lila sank her fangs into his flesh. In an instant, the soldier melted into a pool of murky water.

Silence.

Rosie could only stare at the puddle that used to be a soldier. "Whoa," she said.

Lila and Kimmie hurried back to the stairs. At the top, a large open chamber opened up that was ordained with large statues, potted plants, and an aquarium that covered the entire back wall. Hallways with countless doors shot off in many directions.

Maddy whined as she pointed down the many halls. "Rosie, we'll never find the King and Queen."

Rosie's shoulders weighed heavy with the pressure of the battle. Which way do they go? She did not have time to decide before a wave of stench filled the chamber, and two soldiers jumped from a dark hallway.

Rosie fumbled with the lid on her last Diet Coke, so Giblet and Goblet jumped in front of the soldiers. They smacked their tails in unison three times and a shower of sparks erupted. One soldier's arms and legs turned into twigs that could not support his body weight as he crumbled to the floor. The Larmox wiggled their worm-like bodies onto his head and tugged on his black ear hair.

Sammy waved his top hat like he was riding a bull at the rodeo. "Yee-haaa!"

Lila and Kimmie sank their fangs into the Plyrim soldier and he splashed onto the floor in a puddle.

The second soldier glared at Rosie and her hands shook as she plopped the Mentos into her Diet Coke. She pointed it toward his slimy, squished face and . . . SHE MISSED!

The soldier dodged the bomb and he rushed toward Rosie with hooked claws. Tiki flew through the air and dove into the aquarium. He paddled to lightning speed and launched toward the soldier knocking him off of his feet. The rest of the team pounced. Nugget and Sid grabbed his the legs, while Spike and the boys jabbed in their toxic poison.

The soldier lay lifeless and they all breathed a sigh of relief.

116

Giblet patted Tiki on the back for his water heroics, Nugget and Sid nuzzled their muzzles, and Spike and the Larmox rolled around with joy.

Rosie wished it was over, but they still had to face the Fligarian. She looked around and scratched her chin. "Why don't we go down the hallway that the soldiers came from? I bet they were guarding the Fligarian!"

Lucy grinned. "That's why you're the smart one and I'm the pretty one."

At the end of the dark hallway, a solitary door stood tall and frightening in the silence. Xena examined the door. "I can't see past. It's pitch black on the other side." Xena whirred into circles of anxiety.

Rosie knew they had to go inside, but it was going to be dangerous. She knelt down to Maddy. "Maddy, I want you to wait out here so you're safe."

Maddy stood tall and peered deep into Rosie's eyes. "I can do it, Rosie. I want to help."

Rosie paused. Without Maddy, they would have never found Giblet and Princess Nilly. And it was Maddy that sang the song to free the Ackly vines guarding the prisoners. Rosie nodded her head and smiled. "You're right, Maddy. You CAN do this." She gave Maddy a high five.

Nathan stepped forward. "We have to go for it," he said. "The King and Queen must be on the other side of this door. We have weapons left, but we'll have to fight in the dark."

"I can help." Kimmie slithered forward. In the dim hallway, they could see her producing a green light.

"Kimmie, I forgot that you glow in the dark!" Rosie was elated. "Can you lead us into the room?"

"I'm ready. Lila and I have been ready to defend Grymballia since those smelly soldiers arrived. Let me at them." Her grin glowed.

Rosie, Nathan, and Lucy stood side by side at the heavy door, and with a nod, they thrust it open. The entire team rushed inside the room with Kimmie leading with her neon glow.

An enormous room filled with gemstones, paintings, plants, and statues stretched forever into the distance. An enormous, fluffy bed sat in one corner, and on the bed was the King and Queen with their wings tied together. Their mouths were bound shut with blades of grass and their wide eyes pleaded as they nodded frantically toward the opposite corner.

The Fligarian moved toward them slowly with steam spewing from his nostrils.

Rosie gasped and Lucy screamed. The Fligarian's head skimmed the arched ceilings and Rosie craned her neck to look up into its disgusting face. The head was like an iguana with large nostrils pouring steam to remind her of the fire inside. Black scales covered the skin and long pointed spines covering every inch of its body. The Fligarian feet were spiked with nails that could shred all of Grymballia, but Rosie was horrified to glimpse his eyes. Piercing fire-red eyes with a slit black pupil glared from above as the Fligarian grinned with a mouth full of jagged teeth.

Rosie didn't know if she should hide or fight. How could they beat this monster? She glanced toward the King and Queen bound to the bed and found renewed strength. She clenched her teeth. "Bring it on, smoke-breath."

In response, the Fligarian inhaled with a whirlwind and shot out a burst of fire. Rosie grabbed Maddy and Nugget and threw them out of danger. Everyone scrambled to dodge the flames.

"We need to attack!" Rosie shouted. "It's our only hope!"

A chorus of screams and battle cries arose from the Grymballians. Rosie had no more Mentos or Coke so she helped Lucy with her weapon. They pulled off the lid and Lucy threw in her Mentos in addition to the whole box of Mentos that Rosie had left in her pocket.

The eruption was fantastic! It hit the Fligarian right as it was opening its jaws to torch their friends. It choked and sputtered while Priscilla worked her web magic. Her pink ponytails flew as she stretched her web from the bedposts to the leg of a dresser. She formed a wall of web. Priscilla yelled up to the winged Grymballians.

"Glaperia, Larmox, Tiki, and Princess Nilly; force the Fligarian toward the web!" Priscilla's glittery boots waved as she worked fast.

Under the glow of Kimmie, the winged Grymballians flew into the face of the Fligarian and dodged his gnashing teeth and swinging claws. The Fligarian swatted the air as he walked backward. Rosie looked to Nathan.

"Look!" She said as she pointed.

One more step, and the Fligarian would trip over the web wall. Lucy and Rosie charged in to help. They kicked his scaly shins while Nugget and Sid nibbled his toes. Nathan rushed the Fligarian but was knocked to the ground by his spiked tail. The Fligarian huffed and Rosie knew he was about to spew fire.

Out of the corner, Maddy rushed over and scooped up Nathan's Diet Coke. Just as the Fligarian opened his mouth to shoot fire, Maddy dropped in the Mentos and fired!

The explosion extinguished the fire, and Maddy bounced up and down giggling. The flustered Fligarian stumbled back another step and then tripped over Priscilla's solid wall of web. The massive body tumbled to the ground and the whole castle shook.

Grymballians surrounded the fallen Fligarian as he tried to get back on his feet. They used their talents and weapons as the Fligarian roared in frustration. His nostrils started to steam, and he was building to fire again.

Rosie moved to action with little thought as she grabbed Nathan's backpack and pulled out the two thermoses of liquid nitrogen. Nathan hurried to her side and grabbed one of the containers. Even though the frosty thermos froze Rosie's hands, she opened the top and fastened the spray nozzle onto the sizzling liquid.

The Fligarian steamed and opened his mouth to fire. "Fire, Nathan!"

They simultaneously shot liquid nitrogen into the mouth of the Fligarian. The fire extinguished on contact and the Fligrian's mouth started to turn white with ice. Rosie continued to spray and empty her bottle despite numb, frozen fingers. The Fligarian's entire head froze and he became a solid ice sculpture of scales and fangs. When the canisters were emptied, Rosie and Nathan watched in awe as the Fligrian's frozen head started to fall toward the floor.

"Everyone get back!" Rosie shouted and pointed toward the wall.

The frozen Fligarian fell to the marble floor and the ice-monster shattered into a thousand pieces.

The Fligarian was dead.

The room filled with cheers, hugs, and dancing.

Maddy squealed. "Rosie, you saved Grymballia!"

Rosie pulled her sister close. "No, WE ALL saved Grymballia."

Chapter 17

The King and Queen rubbed their tiny wrists where they had been tied and then launched into the air over the bed to stretch their wings. They hovered above their rescuers.

"Princess Nilly, you saved us and have proven to be an excellent leader. You found Rosie and her friends from the outside world and saved Grymballia." The King and Queen flew toward the doorway. "We must hurry to our thrones to banish the Plyrim soldiers forever and restore peace."

The Grymballians looked upon the royal King and Queen with reverence. The King looked an older version of Princess Nilly with the same twig-like arms and petite body. Pine needles scattered his head and jutted out from under his royal crown. His eyes glowed like yellow sunshine.

The Queen's beauty filled the room as golden maple leaves covered her head under a tiara of gold. Her sunlit eyes warmed the room as she gazed upon Princess Nilly and embraced her daughter. "I'm so proud, Nilly."

Rosie and her team bowed to the King and Queen.

The King said, "Thank you, friends. Let's take back Grymballia."

Applause and cheers erupted.

"We must move carefully, your Highness," Rosie said. "We've battled three soldiers in the dungeon and five more between here and the throne. We don't know how many remain."

The King smiled at Rosie. "I will follow you and your noble team."

Rosie filled with pride and cherished the many eyes looking for her leadership. "Let's move," she said.

Rosie kept Maddy close but knew her sister was braver than she had ever imagined. Nugget kept close to Sid. They stepped over shattered Fligarian parts as they exited the chamber. Patsy scanned the door and motioned that all was clear. Once in the hallway, Kimmie relaxed her neon glow.

"Kimmie, we couldn't have done it without your light," Rosie said

"Thanks, Rosie." Kimmie blushed and slithered back by Lila.

Near the aquarium, one soldier remained sleeping next to his fellow puddle of water. No other Plyrim soldiers lurked near.

"We'll scan the main entrance below," Dina and Nina flitted overhead.

They zipped ahead with wings beating rapidly as they hovered at the top of the stairs. From that point of view, they could see in all directions.

"The fallen soldiers have not moved, and I don't see any soldiers on patrol below, but two soldiers guard the front entrance outside of the castle." Trina reported.

Rosie surveyed her crew. They had no more weapons except for the magic of her Grymballian crew. The goal was to get the King and Queen to their thrones without alarming the soldiers outside. Rosie whispered, "Okay, let go."

As they rushed down the stairs, the mummified soldier wrapped in Priscilla's web squirmed and grunted. Another soldier remained asleep next to a puddle of soldier water. Maddy squeezed Rosie's hand.

Once down the stairs, the King and Queen hurried toward their thrones while the others stood guard. The flowering sun overhead welcomed them home. The Queen's delicate wings flitted gracefully next to her sunflower throne. The King hovered over his mushroom throne and turned to his people. "Friends of Grymballia, you saved our world from destruction. By returning us to our thrones, Plyrim will fall and the soldiers will be banished. We must remember why this happened. Plyrim destroyed their world by disrespecting Mother Nature. We must preserve our water, save our trees, and use our energy wisely to keep Grymballia thriving." The King's voice boomed across the castle.

The King and Queen grasped hands and suddenly the overhead sun bloomed with flowers opening wide to fill the lobby with the sweet scent of peace. The King and Queen floated down toward their thrones.

"HALT!" The growl of a Plyrim soldier approached from behind. His stench drowned out the smell of flowers.

The soldier hurried in from the outside and rushed toward the King and Queen.

"Stop him!" Lucy screamed.

The King and Queen looked panicked as they hovered inches from their throne.

Rosie's team attacked, and the soldier had no clue that he was surrounded. Nugget and Sid gnawed on each leg, while Lila and Kimmie coiled and then bounced off of his face. Tiki flipper-slapped his nose until they forced him to fall on the ground. Spike, Rufus, and Dennis jumped on top of him, while Goblet and Giblet prepared to do magic. Priscilla wove a web of steel to tie his feet together with while Larmox and Glaperia swarmed his face to blind him. Everyone worked together as a team . . . and then it happened.

The King and Queen rested onto their thrones.

Radiant sunlight blazed into the room and Rosie shielded her eyes. She watched in awe as the Plyrim soldier vanished from under the pile of Grymballians. She scanned the room and every last soldier was gone. Grymballia was free.

Silence filled the room and everyone looked to each other for confirmation – it was over. Arms, flippers, wings, and tails flew into the air with cheers of joy. Rosie ran to the dungeon door and opened it wide. The Grymballians flowed out from the prison below to join the celebration.

Rosie noticed a serious conversation between Princess Nilly and her royal parents. The King clapped his hands and his gold crown glowed. The Ackly vines crawled out of the dungeon into a sunlit world that they had not seen for years. The Princess was true to her word and freed the Ackly vines forever.

The vines paused at the front doors and everyone cheered their freedom. Nathan opened the front door, and the vines rushed into the fresh air. They wrapped around a statue of Mother Nature and basked in the sunlight.

When the celebration quieted, Rosie knew they must return home. Princess Nilly cried tears of thanks and flew circles around their heads.

"We were an awesome team." Rosie looked to all of her new friends. "We couldn't have done it alone. You have a beautiful world and the Grymballians are wonderful." Rosie's chest clenched as she looked into the many eyes of the Glaperia and as Tiki waved his flipper. She hated to leave.

"You'll always special, Rosie. You and your friends are always welcome in Grymballia." Princess Nilly landed on Rosie's shoulder.

Warm tears ran down Rosie's face. "Thank you, Princess Nilly. I would love to visit again someday."

The sun dipped low on the horizon, and Rosie knew they must get home before her parents searched the forest.

"Princess Nilly, we have to go."

Maddy stood side by side with Giblet holding his hand with a tear-stained face. They had a special friendship.

"I have something for you as a token of thanks." Princess Nilly handed Rosie a golden acorn necklace. "When you want to return to Grymballia, bring this to the cave. Giblet will come escort you back to our world."

Rosie grasped the acorn against her chest. The princess sang a beautiful song, and the Blim Birds arrived. Giblet would ride to the caves to escort them home, but first the entire Grymballian team surrounded them. Sammy tipped his hat as he stood next to the other Larmox, and Tiki waved his flippers. Patsy and Jojo flapped their wings, and Rosie loved their new friendship. Lila and Kimmie bounced a good-bye, while Priscilla spun a web into the shape of a heart. The Plumpians puffed out their spikes as a tribute to our battles, while Xena, Trina, Dina, and Nina wished them well. Nugget and Sid licked noses to say good-bye. They boarded their Blim Birds for the journey home.

As they flew over Grymballia, Rosie looked down upon Grymballia with satisfaction that they had saved a beautiful world. Maddy squealed as the Blim Bird glided through a rainbow and Lucy waved her fist in the air.

Maddy turned to Rosie. "When will we come back to visit our friends? I didn't get to see Giblet's house."

Rosie rubbed the golden acorn hanging from her neck and didn't know the answer. If they repeatedly returned to Grymballia, someone might discover their secret world. "I don't know, Maddy. We need to keep Grymballia a secret so that nobody hurts Giblet and our friends."

"Who would hurt Giblet?" Maddy gasped.

"We live in a different world, Maddy," Rosie said.

Maddy leaned against Rosie's back and her warmth and innocence made Rosie smile. Her sister wasn't so bad after all.

The Blim Birds landed next to the cave, and Rosie knew they would never look at a rainbow the same.

Inside the cave, Maddy knelt next to Giblet. "I'm going to miss you so much, Giblet. You're my best friend. I wish you could come to my house for another tea party."

Giblet's head dropped and he grabbed Maddy's hand. "I will always miss you, Maddy."

"I have something for you." Maddy reached into her pocket and pulled out a Dum-Dum. "It's the last one. I saved it for you."

Large tears fell down Giblet's face. He reached for the Dum-Dum and held it close. He hopped onto a rock and kissed Maddy's cheek.

"Oh, Giblet!" Maddy giggled with delight.

Giblet drew a pattern on the cave's dirt floor, and then started to chant in a low hum. Rosie noticed that the words to return home were different:

Land of the Earth, we must leave you now,
We keep all your secrets, we solemnly vow.
Nature's our friend to never neglect
Grymballia we leave but always protect.

He slapped his tail three times and sparks shot out as the portal glowed. Giblet's face faded away as they passed through the portal.

They landed in the familiar cave by the bluff, and Nugget ran around sniffing at her home territory. Outside the cave, the sun dipped low.

"We better hurry home," Rosie said.

Initially they hiked in silence. The warm air welcomed them home, and Rosie felt taller and stronger as she looked around. She was part of something important and her love of the environment played a roll in saving Grymballia. She put her arm around Maddy and gently tugged on her pigtail. "I'm proud of you, Maddy. You did great today."

Maddy stopped and looked up at Rosie. "Really, Rosie? You're proud of me?" Maddy threw her arms around Rosie's waist.

Rosie warmed inside and decided they needed to hug more often.

Nathan dragged his feet. Rosie said, "Nathan, I bet you're tired. You've been back and forth to Grymballia twice today to get the weapons."

He leaned against a tree to rest. "I'm exhausted. When I came back, I rode my bike to the gas station for the Coke and Mentos and then hurried to the school for the liquid nitrogen. I hope Mr. Barclay doesn't notice that some is missing." He rubbed his legs and started to hike again.

"Did Mom notice you?" Rosie asked.

"I don't think so. I snuck around the side of the house so she couldn't see me."

As they reached the end of the trail, Rosie's house loomed in the clearing. She didn't want the adventure to end.

"Grymballia will be our secret so we can protect our friends." Rosie said as Nathan, Lucy, and Maddy nodded.

In the back yard, Nugget barked and charged for the back door. Rosie's dad grilled on the back patio.

"You're back! We almost sent out a search party." He laughed. "Did you have any adventures in the forest?"

She glanced at her friends. "Not really, Dad." Rosie rubbed the golden acorn as she envisioned Grymballia and the epic battle in her mind. She grinned. "We'll always have our secrets in the forest."

Epilogue

Rosie sat on her bed as the breeze from her open window carried the smell of fresh pine. She stared at the ocean mural on her wall, and couldn't help but think about Tiki swimming among the sea turtles. The acorn necklace rested on her chest as a constant reminder of Grymballia. She heard Maddy singing from her room.

Bright sunshine and rainbows,
Playing children to and fro,
Flowers, trees, and green grass,
Springtime's here at last.

Rosie suspected Maddy sang this song when she thought about Grymballia. She and Maddy had a new friendship and Rosie played tea party with Baby Annabelle or pushed her on the swing. Their parents were baffled by Maddy's newfound love of pears.

After Grymballia, Lucy seemed less like a diva princess. Her father called her "Noodle" like when she was a baby, and she convinced him to start a recycling program on her block. The most exciting news is that she WAS going to Disney World with her dad next month.

Nathan asked Rosie to be his partner for the science fair and Rosie felt she might explode with excitement. They were meeting next week to develop a project to raise awareness about pollution in the river. His parents were starting to understand that maybe he had a gift for science and engineering with no interest in art. Nathan's parents enrolled him in chemistry camp the next week.

Nugget missed Sid and often wandered toward the forest trail to sniff. It had been three weeks since their adventure, and Rosie missed Grymballia. She paged through her journal often to remember Grymballia and her new friends. A breeze blew her hair as she stared toward out her window toward her backyard forest filled with secrets. The crickets chirped and cicadas hummed, and then Rosie heard Princess Nilly's song:

Be at peace, Miss Rosie
We miss you too.
Go forth with your life
We need that of you.
Grymballia is safe now,
Thanks to you and your friends.
Be at peace, Miss Rosie,
We will see you again.

Princess Nilly continued to sing, and her voice lulled Rosie to sleep as she grasped her acorn and dreamed of Grymballia.

ACKNOWLEDGMENTS

Thank you for reading the second edition of SAVING THE WORLD. After its initial production years ago, I realized that writing pulled me to imaginary worlds that I could craft to entertain children. I also realized that I had more to learn about the craft of writing. I earned my MFA in Creative Writing for Children and have multiple pages filled with imagination and adventure. I hope to continue writing and sharing them all with you!

I could not dedicate myself to writing without the help and support of my family and friends. Being a physician, I had to make tough choices to find a balance between writing, mommying, and my medical career. Thank you to my husband, Ryan, and my children for supporting my dreams. Thank you to my parents for always being my biggest cheerleader. Thank you Casey Feller for being my personal graphic designer – couldn't have done it without you. Kimmie Van Scyoc encouraged me to write my first book, and I haven't looked back.

Coming Soon!!

Return to Grymballia

The second book of the Grymballia series Visit: patmccawauthor@gmail.com for release date and information.

CPSIA information can be obtained
at www.ICGtesting.com
Printed in the USA
FFHW022217281118
49664214-54078FF